Springer Series in Advanced Microelectronics

Volume 15

The *Springer Series in Advanced Microelectronics* provides systematic information on all the topics relevant for the design, processing, and manufacturing of microelectronic devices. The books, each prepared by leading researchers or engineers in their fields, cover the basic and advanced aspects of topics such as wafer processing, materials, device design, device technologies, circuit design, VLSI implementation, and sub-system technology. The series forms a bridge between physics and engineering, therefore the volumes will appeal to practicing engineers as well as research scientists.

Kaiying Wang

Photoresist Technology in Microsystems: Principles, Processes and Applications

Springer

Kaiying Wang
Department of Microsystems
University of South-Eastern Norway
Horten, Norway

ISSN 1437-0387 ISSN 2197-6643 (electronic)
Springer Series in Advanced Microelectronics
ISBN 978-981-95-1605-6 ISBN 978-981-95-1606-3 (eBook)
https://doi.org/10.1007/978-981-95-1606-3

This Springer imprint is published by the registered company Springer Nature Singapore Pte Ltd.
The registered company address is: 152 Beach Road, #21-01/04 Gateway East, Singapore 189721, Singapore

Preface

Photoresist technology stands as the cornerstone of modern microelectronics and microsystem fabrication, playing a pivotal role in the development of semiconductor chips and other high-precision microstructures. Over the last several decades, the evolution of photoresist materials and processing techniques has enabled the creation of smaller, more efficient, and highly functional devices. This book aims to provide a comprehensive exploration of the principles, processes, and applications of photoresist technology within the field of microsystems.

The motivation behind this book stems from the growing complexity of microsystem design and manufacturing, which demands an in-depth understanding of both the materials and techniques involved. As microsystem-based applications expand into fields such as healthcare, automotive, telecommunications, and environmental monitoring, the need for advanced photoresist materials and precision processing has become more pressing than ever.

This book is organized to provide a structured and thorough understanding of photoresist technology. Beginning with an introduction to the fundamental chemistry of photoresist materials, the book progresses through detailed chapters on fabrication processes, patterning techniques, and the challenges faced in manufacturing. Special focus is given to advanced photoresist materials, high-resolution lithography, and innovative processing techniques. The final chapters address the latest research trends, including AI-driven lithography optimization, sustainable photoresist materials, and future perspectives in flexible and wearable microsystems.

This work reflects the collective efforts of the students and contributors whose dedication and expertise have made this book possible. It is intended to serve as a valuable resource for researchers, engineers, and students working in the fields of microsystem engineering and advanced materials.

We hope this book provides you with both practical insights and theoretical foundations in photoresist technology, helping to drive further innovation and discovery in the field of microsystems.

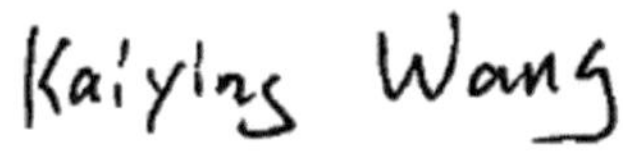

Horten, Norway
July 2025

Prof. Kaiying Wang

Acknowledgments

We would like to express our deepest gratitude to the 26 EU Erasmus and USN Master degree students who took the course *SSI Design Lab II—Manufacturing Processes for Smart Systems (JMN455)* and *Microfabrication Technology (MFA4000)* at the University of South-Eastern Norway. Their hard work, dedication, and insightful contributions have been instrumental in the successful completion of this book.

I am especially grateful to Dr. Tao Feng, Kun Zhao (Chaps. 2 and 3), Zhuohao Liu (Chaps. 4–6), and Yu Cui (Chaps. 7 and 8) at Taiyuan University of Technology for their outstanding efforts in editing all chapters of the final version and for securing the copyrights. Their commitment to quality and scholarly integrity has been invaluable.

The contributions of the Erasmus students to individual chapters are acknowledged as follows:

- Chapter 2: Adnan, Md. Sahib Alam Provat, Md. Shadman Shakif Bhuiyan, Vaibhav Ashok Kumar
- Chapter 3: Gabriel Andreas Xavier Facusse Zepeda, Asma Mohiuddin, Hesham Mohamed Ellabban, Laura Daniela Gallego García
- Chapter 4: Fjolla Ahmetaj, Pragya Mallick, Hoàng Vũ Nguyễn
- Chapter 5: Ikenna Idu Henry, Ismael Musa, Maira Ehsan Mughal, Natnael Masresha Zerihun
- Chapter 6: Sanket Sandeep Deshmane, Ali Ghasemibousjin, Džeina Dombrava, Pegah Haghshomar
- Chapter 7: Nico Wichmann, Rodrigo Zamagno Medeiros, Lina Khaled Alkerdi, Muhammad Ukasha
- Chapter 8: Amir Abdelkareem Osman Tahir, Ahmed Abdullah Nasir, Mohammad Toufan, Ziwei Liu

Their enthusiasm and commitment to this project have been truly commendable. We are grateful for their valuable insights and volunteer contributions, which have significantly enriched the quality and depth of this book.

I would also like to extend my sincere appreciation to all the highly intelligent students from around the world who joined the programme and studied at the University of South-Eastern Norway as part of their mandatory course. Your hard work, innovative thinking, and global perspectives have brought immense value to the learning environment and have played a significant role in the success of this book project.

As the main professor responsible for this course, I would like to give my highest appraisal to these students for their hard work and their very positive attitude throughout the course. Their professionalism, dedication, and willingness to engage deeply with the subject matter have made this project a remarkable success.

Special thanks also go to the academic and technical staff at the University of South-Eastern Norway for their guidance and support throughout the development of this book.

This book is a testament to the spirit of collaboration and shared knowledge. Thank you to everyone who contributed to making this work a success.

Kaiying Wang

Horten, Norway Prof. Kaiying Wang

Contents

Chapter 1
Introduction to Photoresist Technology

1.1 Overview of Photoresist Technology in Microsystems

Photoresist technology plays a fundamental role in the fabrication of modern electronic and microscale systems. It is essential for the production of integrated circuits (ICs), sensors, and microelectromechanical systems (MEMS), all of which form the backbone of today's digital and smart technologies. Photoresists are light-sensitive polymeric materials that undergo chemical changes when exposed to specific wavelengths of radiation—typically ultraviolet (UV) light. This property enables them to serve as precise patterning tools during photolithography, allowing engineers to define intricate micro- and nanoscale features on various substrates. In essence, photoresists act as high-resolution stencils that facilitate the controlled transfer of design patterns onto surfaces, making it possible to construct complex microstructures with exceptional accuracy and scalability.

A photoresist is a light-sensitive polymeric material that undergoes a chemical transformation when exposed to specific wavelengths of radiation, most commonly in the ultraviolet (UV) range. This photosensitivity enables the selective dissolution or retention of the photoresist during the development process, based on the light exposure pattern. As a result, intricate patterns can be accurately transferred onto the surface of a substrate, forming the basis of high-resolution photolithographic techniques essential in microfabrication [1–4].

The significance of photoresists lies in their capacity to produce sub-micron precision patterns, which are critical for the continued miniaturization and enhanced performance of integrated circuits (ICs), MEMS, and other micro- and nanoscale devices. Figure 1.1 illustrates a manual coating process using the commercial photoresist AZ 4562 on a 4-inch silicon wafer for MEMS fabrication at the Norwegian Infrastructure for Micro- and Nanofabrication (NorFab) facility at the University of South-Eastern Norway (USN).

K. Wang, *Photoresist Technology in Microsystems: Principles, Processes and Applications*, Springer Series in Advanced Microelectronics 15,
https://doi.org/10.1007/978-981-95-1606-3_1

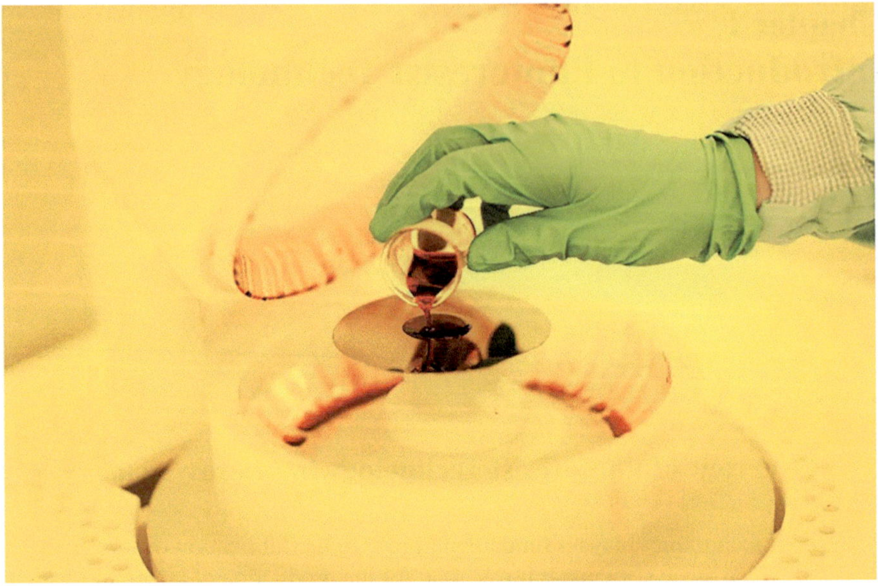

Fig. 1.1 Application of photoresist on silicon wafer at NorFab USN-labs

Beyond their role in pattern definition, photoresists are instrumental in ensuring dimensional control, process reproducibility, and integration compatibility in high-volume manufacturing. Their performance directly influences yield, scalability, and device reliability. Indeed, the development of modern semiconductor and microsystem technologies would not have been possible without the continuous innovation and refinement of photoresist materials and processing techniques.

Brief Introduction to Their Role in Micro and Nanotechnology

In micro- and nanotechnology, photoresists serve as the foundational materials in photolithography, a core technique used to define precise patterns on substrates. By coating a substrate with photoresist and exposing it to light through a photomask, photolithography enables the transfer of complex designs with high spatial resolution. This method is essential in the fabrication of integrated circuits (ICs), where feature density and design complexity continue to increase with each successive technology node.

Beyond traditional silicon-based microelectronics, photoresists are employed in the patterning of a diverse range of functional materials—including polymers, metals, and ceramics—to support the fabrication of advanced microsystems. These include microfluidic devices, biomedical implants, optical components, and sensors. In nanotechnology, photoresists capable of producing features at the nanoscale are indispensable for emerging applications such as quantum computing and nanoelectronics [1, 2].

The versatility of photoresists is further demonstrated in additive manufacturing, particularly in stereolithography. Here, photoresists act as photosensitive resins that enable the layer-by-layer construction of complex 3D microstructures, often impossible to achieve using traditional subtractive methods.

Applications in Micro- and Nanotechnology

In microtechnology, photoresists commonly serve as structural templates or sacrificial layers for etching, material deposition, and ion implantation processes. Notable examples include:

- MEMS devices: Negative-tone resists such as SU-8 are widely used to fabricate high-aspect-ratio structures, such as accelerometer proof masses and microcantilevers.
- Microfluidics: Molds made from photoresists like PDMS are used to define microchannels for lab-on-a-chip diagnostic platforms.

In nanotechnology, the development of advanced resists has enabled critical breakthroughs in pattern resolution and material compatibility:

- Extreme Ultraviolet (EUV) Lithography: Chemically amplified resists (CARs) enable < 10 nm patterning required for 3 nm-node semiconductor manufacturing.
- Nanoimprint Lithography (NIL): UV-curable resists allow replication of sub-20 nm nanostructures for applications in photonics and metamaterials.
- Directed Self-Assembly (DSA): Photoresist-based templates guide block copolymer alignment, enabling the formation of periodic sub-5 nm patterns.

Photoresists also serve as cross-disciplinary enablers in fields such as:

- Biomedical Engineering: Bioresorbable photoresists are used to create temporary implants that safely dissolve in the body.
- Energy Systems: Photoresist-derived structures improve the architecture of solar cells and battery electrodes by enhancing interfacial surface area and charge transport efficiency.

Evolution and Future Directions of Photoresist Technology

The development of photoresist materials has closely followed the trajectory of device miniaturization in microelectronics. Early photoresists consisted of simple, single-component materials; modern resists are complex, multicomponent systems specifically engineered for stringent process demands. Innovations such as chemically amplified resists (CARs) have significantly advanced both sensitivity and resolution. More recently, metal-containing resist films have been introduced for EUV lithography to improve line edge definition and etch resistance [5–8].

Looking ahead, the future of photoresist technology is intertwined with continued efforts to overcome the physical limitations of conventional photolithography, including diffraction and stochastic effects. Alternative high-resolution patterning techniques—such as electron beam lithography, nanoimprint lithography, and multi-beam systems—are being actively explored to push beyond current resolution thresholds.

Recent trends in photoresist research are characterized by a focus on:

- Sustainability: Development of solvent-free, water-based, and bio-derived resists to reduce environmental and health impacts.
- Multifunctionality: Integration of nanoparticles (e.g., conductive or magnetic) into resist matrices to create functional hybrid materials.
- AI-driven process optimization: Application of machine learning models to predict resist behavior, pattern fidelity, and defect formation, thereby accelerating development and reducing cost.

In summary, photoresist technology remains central to the advancement of micro- and nanofabrication. Its continued evolution is not only essential for meeting the demands of smaller, faster, and more energy-efficient devices but also for enabling the next generation of innovations across fields as diverse as quantum computing, biosensing, flexible electronics, and energy storage.

1.2 Evolution of Microsystems and Their Applications

The evolution of microsystem technology, from its conceptual beginnings to its current role in advanced industrial applications, represents a compelling narrative of innovation, precision engineering, and progressive miniaturization. Central to this transformation is the advancement of photoresist technology, which has facilitated the fabrication of increasingly complex, miniaturized, and multifunctional devices. This section outlines the historical development of microsystems and highlights their diverse applications across major technological domains.

Historical Progression of Microsystem Technologies

The origins of microsystem technology can be traced back to the late 1950s, when Jack Kilby and Robert Noyce independently developed the integrated circuit (IC)—a seminal innovation that laid the foundation for miniaturized electronic systems [9]. This milestone, which enabled the integration of multiple electronic components on a single silicon chip, catalyzed the explosive growth of the semiconductor industry in the following decades. The subsequent refinement of photolithography, etching, and deposition techniques allowed for the scalable manufacturing of microdevices with increasing complexity and shrinking feature sizes.

In the 1970s and 1980s, the field evolved further with the emergence of Micro-Electro-Mechanical Systems (MEMS)—systems that integrate mechanical structures, sensors, actuators, and electronics on a single chip. Enabled by silicon micromachining (including both wet and dry etching), this era saw the development of the first MEMS pressure sensors for automotive applications, such as manifold pressure sensing. These advances demonstrated the viability of MEMS as a platform for compact, high-performance mechanical-electronic integration.

The 1990s to early 2000s marked the commercial breakthrough of MEMS technologies. Devices such as accelerometers (e.g., the ADXL series by Analog Devices), inkjet printer nozzles, and micromirrors for Digital Light Processing (DLP) entered mainstream markets. Simultaneously, lab-on-a-chip systems emerged, leveraging microfluidic MEMS to perform biomedical assays and diagnostics with high speed and portability. The ability to manipulate fluids at the microscale paved the way for applications in clinical testing, drug delivery, and environmental sensing.

With the onset of the 2010s, microsystem research and manufacturing entered the nanotechnology era. The integration of MEMS with nanostructures gave rise to Nano-Electro-Mechanical Systems (NEMS), capable of achieving atomic-scale motion and ultra-sensitive detection. These developments were underpinned by the use of advanced photoresists and next-generation lithography techniques, including Extreme Ultraviolet (EUV) lithography, which allowed for sub-10 nm feature resolution.

Simultaneously, MEMS technologies became foundational to the Internet of Things (IoT) and wearable electronics. Compact sensors such as gyroscopes, accelerometers, pressure sensors, and microphones became standard components in smartphones, smartwatches, and fitness trackers. The adoption of 3D printing and additive manufacturing further enhanced the fabrication of microsystems, enabling more rapid prototyping and integration of complex geometries with diverse materials.

The evolution of microsystem technologies has profoundly impacted a wide range of industries, leading to transformative advances in product design, performance, and functionality [11–14]. As illustrated in Fig. 1.2, microsystems now play a pivotal role across sectors such as healthcare, automotive, telecommunications, consumer electronics, environmental monitoring, and energy.

Healthcare Industry

In healthcare, microsystems have revolutionized both diagnostic and therapeutic capabilities. Miniaturized sensors now enable continuous monitoring of vital signs, while microfluidic devices facilitate rapid, high-throughput analysis of biological samples—significantly enhancing point-of-care testing. Implantable medical devices, such as pacemakers, insulin pumps, and targeted drug delivery systems, have benefited from MEMS integration, which enhances device reliability, energy efficiency, and patient safety. Furthermore, the proliferation of wearable health monitors equipped with microsensors empowers users to track health metrics such as heart rate, blood oxygen, and movement, supporting real-time health management and early disease detection.

Automotive Industry

Microsystems have become integral to automotive safety, efficiency, and autonomy. MEMS-based inertial sensors underpin airbag deployment systems, anti-lock braking systems (ABS), and electronic stability control (ESC). The rise of Advanced

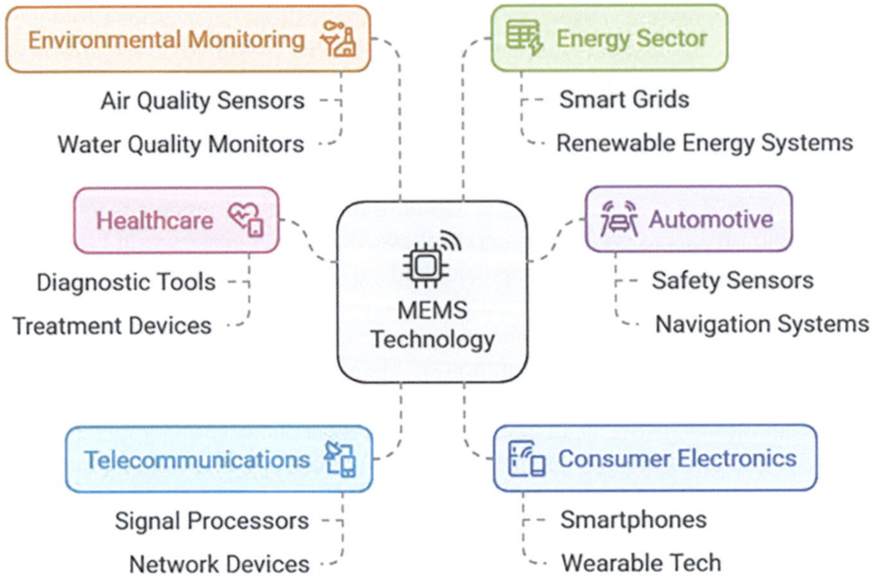

Fig. 1.2 MEMS applications across industries

Driver-Assistance Systems (ADAS) relies heavily on microsensors for functionalities including lane-keeping, collision avoidance, and adaptive cruise control. Additionally, MEMS components embedded in engine management systems enhance combustion control, resulting in improved fuel efficiency and emissions reduction.

Telecommunications Industry

In the telecommunications sector, microsystems have enabled the miniaturization and performance enhancement of wireless communication devices. Radio Frequency (RF) MEMS components—such as tunable capacitors, resonators, and switches—offer lower power consumption, faster switching, and improved signal integrity compared to conventional components. These advantages have supported the development of compact, efficient smartphones and IoT devices. Moreover, micro-optical MEMS have played a crucial role in fiber-optic communication, facilitating the high-speed transmission of large data volumes over long distances with low latency.

Consumer Electronics

Microsystem technologies are ubiquitous in consumer electronics, where they contribute to the enhanced functionality, compactness, and energy efficiency of devices. Smartphones, tablets, smartwatches, and augmented reality (AR)/virtual reality (VR) devices incorporate numerous MEMS components—including accelerometers, gyroscopes, magnetometers, pressure sensors, and microphones—to support intuitive user interfaces, motion tracking, and contextual awareness. Their small footprint and low power consumption have been key enablers of the mobile computing revolution.

Environmental Monitoring

In the environmental domain, microsystems offer real-time monitoring solutions for air, water, and soil quality. Miniaturized chemical and physical sensors are deployed in remote or hazardous environments to detect pollutants, track climate-related variables (e.g., temperature, humidity, CO_2 concentration), and monitor ecological conditions. These technologies support early warning systems, environmental compliance, and data-driven sustainability initiatives.

Energy Sector

Microsystem technologies are also making a significant impact in the energy sector, particularly in the optimization and management of renewable energy systems. MEMS sensors are used to monitor and regulate the performance of solar panels and wind turbines, improving energy harvesting efficiency. Microfluidic platforms are being explored for fuel processing, catalytic conversion, and electrochemical sensing in advanced energy storage systems. In smart grids, microsystems facilitate real-time sensing and control, enabling more reliable and adaptive energy distribution.

The widespread adoption of microsystem technologies across these industries has been made possible by sustained progress in photoresist materials and photolithographic processing techniques. These advancements allow for ever-smaller, more precise, and functionally integrated components. As micro- and nanofabrication technologies continue to evolve, the influence of microsystems is expected to grow, catalyzing new generations of devices that are smarter, more sustainable, and more intimately integrated into daily life.

1.3 Importance of Photoresists in Microsystems Fabrication

Photoresists occupy a central and strategic role in microsystems fabrication, acting as the linchpin of photolithography - the foundational technique that enables the production of virtually all micro- and nanoscale devices. Their importance extends well beyond that of functional materials; photoresists are integral to achieving the precision, scalability, and innovation that define the field of microsystems technology [15–20]. This section explores their core role in photolithography and their broader implications across technological domains.

Photolithography is a multistep process involving the coating of a substrate with a photoresist, exposure to patterned light through a photomask, and subsequent development to define specific geometries. The photoresist governs the fidelity of the transferred pattern, and its properties are crucial for determining resolution, dimensional control, and process compatibility. Critical performance attributes include:

- High photosensitivity
- Strong substrate adhesion
- Chemical and plasma etch resistance
- Thermal and mechanical stability

Photoresists are typically categorized into positive and negative types. In positive resists, exposed regions become more soluble in the developer and are removed; in contrast, negative resists become cross-linked upon exposure and remain after development. The selection of resist type depends on factors such as feature shape, resolution requirements, and post-lithography processes.

The evolution of photoresist materials has directly enabled advances in photolithography. Early photoresists offered limited resolution, suitable only for micron-scale patterning. The introduction of chemically amplified resists (CARs) in the 1980s represented a paradigm shift, allowing for the definition of sub-micron features with high sensitivity and throughput. Today, specialized resists designed for extreme ultraviolet (EUV) lithography support feature sizes well below 10 nm, paving the way for next-generation semiconductors and nanodevices [21].

Precision is a fundamental requirement in microsystems engineering, where the accurate definition of geometries at sub-micron and nanometer scales directly impacts device performance. The resolution capability of a photoresist—defined as the minimum feature size that can be consistently patterned—determines the limits of system miniaturization. In integrated circuit (IC) fabrication, finer features enable greater transistor densities, higher switching speeds, and improved energy efficiency.

Moreover, photoresists facilitate the fabrication of complex three-dimensional microstructures required in microfluidics, optical MEMS, and biomedical devices. For instance, creating high-aspect-ratio microchannels and flexible microvalves demands not only high resolution but also structural integrity and process compatibility—requirements fulfilled by carefully engineered photoresists.

The scalability of microsystems fabrication hinges on the ability to replicate high-resolution patterns uniformly across large wafer surfaces. Photoresists must exhibit excellent uniformity, process stability, and defect tolerance to ensure consistent performance at scale. The transition from 200 to 300 mm wafer platforms in the semiconductor industry, for example, was made possible by significant advancements in photoresist formulation and spin-coating processes.

Modern photoresists support high-throughput lithography systems and advanced overlay control, which are essential for multi-layer device architectures. These innovations have enabled the cost-effective manufacturing of MEMS, sensors, and integrated photonic devices, contributing to the widespread adoption of microsystems in consumer electronics, automotive systems, and industrial sensing platforms.

Innovation is the engine of progress in microsystems technology, and photoresists are among its most dynamic drivers. The continuous development of novel photoresist materials has expanded the possibilities for new patterning techniques. For instance:

- Nanoimprint lithography (NIL) leverages mechanically deformable resists for high-fidelity pattern transfer using physical molds.

- Block copolymer lithography uses photoresist templates for directed self-assembly of sub-5 nm features.
- Hybrid organic–inorganic resists are being developed for enhanced etch resistance and EUV compatibility.

These innovations have enabled the emergence of nanoelectromechanical systems (NEMS), quantum dots, and metamaterials—technologies with transformative potential in fields such as computing, photonics, and biotechnology.

The synergy between advanced photoresists and emerging fabrication platforms such as 3D lithography, direct-write laser patterning, and AI-optimized resist design further underscores the centrality of resist materials in enabling next-generation microsystems.

Conclusion: Photoresists are far more than passive materials in microsystems fabrication—they are active enablers of technological progress. By governing the limits of resolution, ensuring manufacturability at scale, and supporting novel patterning paradigms, photoresists shape the pace and direction of innovation in microsystem engineering. As research continues to extend the frontiers of patterning technologies and materials science, photoresists will remain at the core of efforts to fabricate ever-smaller, faster, and smarter micro- and nanoscale systems.

1.4 Objectives and Scope of the Book

This book provides a comprehensive and in-depth exploration of photoresist technology, offering both fundamental insights and practical knowledge for researchers, engineers, and professionals engaged in microelectronics, nanotechnology, and materials science. By systematically addressing the chemistry, processing techniques, and diverse applications of photoresist materials, the text aims to bridge the gap between scientific theory and industrial implementation. Ultimately, it supports innovation and the advancement of microsystem fabrication by equipping readers with the tools to meet evolving technological demands.

A Resource for Researchers and Engineers

The primary objective of this book is to serve as a foundational reference for the development and application of photoresist technology in the fabrication of micro- and nanoscale devices. It demystifies the complexities of photoresist formulation and photolithographic processes, enabling readers to address fabrication challenges with greater confidence and precision.

For researchers, the book offers a rigorous examination of the molecular design principles and photochemical mechanisms that define photoresist performance. By elucidating how factors such as polymer structure, photoacid generation, and crosslinking reactions influence pattern resolution, sensitivity, and environmental stability, the text provides a framework for developing next-generation photoresists tailored for emerging applications.

For engineers, the book emphasizes the practical aspects of photolithography. It includes detailed discussions of exposure strategies, development protocols, and post-exposure treatments, offering actionable guidance for process optimization in industrial environments. The interplay between chemical formulation and processing conditions is highlighted throughout, enabling process engineers to make informed decisions that enhance device yield, reliability, and manufacturability.

Structure: Chemistry, Processing, and Applications.

To meet the needs of its audience, the book is organized into three major sections: chemistry, processing techniques, and applications.

1. Photoresist Chemistry

This section begins by introducing the principles of photochemistry, explaining how photoresists undergo light-induced chemical transformations to form patterned structures. It categorizes major classes of photoresists—positive, negative, chemically amplified resists (CARs), and metal-containing formulations—highlighting their synthetic pathways, material properties, and functional advantages. Special attention is given to key components such as photoacid generators, resin binders, solvents, and crosslinkers, and how their molecular interactions define resolution limits and environmental robustness. Mechanistic insights are supported by examples and structure–function relationships, allowing readers to understand and tailor material performance for specific use cases.

2. Photolithographic Processing Techniques

This section provides a thorough guide to the fabrication workflow, from substrate preparation and spin coating, to exposure, development, and post-exposure baking. Emphasis is placed on the control of critical parameters—such as exposure dose, ambient humidity, thermal profiles, and developer chemistry—that impact feature fidelity and reproducibility. Advanced lithographic techniques are also discussed, including:

- Immersion lithography for sub-wavelength patterning
- Multiple patterning for high-density integration
- Directed self-assembly (DSA) for nanostructure formation

The section concludes with an analysis of common failure modes and troubleshooting approaches, providing engineers with practical tools for diagnosing and solving processing issues.

3. Applications Across Emerging Technologies

The final section explores how photoresists enable innovations across a broad range of technological domains:

- Semiconductor Industry: Photoresists are foundational to the patterning of integrated circuits (ICs), facilitating the production of transistors with ever-shrinking feature sizes. Their performance directly impacts device speed, energy efficiency, and packing density.

- MicroElectroMechanical Systems (MEMS): The book explains how thick and high-aspect-ratio photoresists (e.g., SU-8) enable the fabrication of microsensors, actuators, and microfluidic components on a single chip.
- Biomedical Engineering: Applications include the fabrication of implantable devices, biosensors, and controlled-release drug delivery systems, with emphasis on biocompatibility, miniaturization, and integration with living tissue.
- Optical Communications: Photoresists are used to form waveguides, gratings, and microlenses, which are vital for photonics-based data transmission in high-bandwidth networks.
- Energy Technologies: The text discusses the role of photoresists in fabricating solar cell structures, fuel cell components, and energy storage systems, where fine feature control improves efficiency and performance.

Conclusion: This book offers a holistic and authoritative perspective on photoresist technology. By covering its chemical foundations, processing intricacies, and real-world applications, it aims to serve as both a technical manual and an inspiration for innovation. Whether you are a seasoned researcher seeking to design the next generation of photoresist materials, or a process engineer working to optimize high-volume manufacturing, this book provides the knowledge and context needed to contribute meaningfully to the evolving field of micro- and nanofabrication.

References

1. M.J. Madou, *Manufacturing Techniques for Microfabrication and Nanotechnology*, 2nd edn. (CRC Press, Boca Raton, 2011)
2. T.R. Hsu, *MEMS and Microsystems: Design, Manufacture, and Nanoscale Engineering*, 2nd edn. (John Wiley & Sons, Hoboken, 2008)
3. C.G. Willson, M.D. Stewart, *"Photoresists", in Encyclopedia of Materials: Science and Technology*, 2nd edn. (Elsevier, Amsterdam, 2001), pp.6973–6977
4. R. Schwalm, *"Photoinitiators and Photopolymerization", in Encyclopedia of Materials: Science and Technology* (Elsevier, Amsterdam, 2001), pp.6946–6951
5. M. Hassaan, U. Saleem, A. Singh, A.J. Haque, K. Wang, Materials **17**, 2552 (2024)
6. M.W. Hasan, L. Deeb, S. Kumaniaev, C. Wei, K. Wang, Micromachines **15**, 1122 (2024)
7. W. Wei, J. Liu, H. Li, Q. Mu, X. Liu, Prog. Chem. **26**, 1867 (2014). https://doi.org/10.7536/PC140729
8. Y. Tang, *Highlights in Science, Engineering and Technology—WPEMD* **29**, (2023)
9. Wikipedia contributors, Invention of the integrated circuit, accessed February 18, 2023: https://en.wikipedia.org/wiki/Invention_of_the_integrated_circuit
10. Wikipedia contributors, "MEMS," accessed February 18, 2023: https://en.wikipedia.org/wiki/MEMS
11. Fortune Business Insights, Electronic chemicals and materials market size & share [2028], (2022), accessed February 18, 2023.
12. IMARC, Electronic chemicals and materials market share, size, growth, opportunity and forecast 2022–2027, (2022), accessed February 18, 2023
13. IEEE IRDS. Line Edge Roughness (LER) White Paper, (2022), accessed February 17, 2023: https://irds.ieee.org/images/files/pdf/2022/LER_white_paper.pdf
14. Irresistible Materials Ltd., Next generation photoresist technology for microchips," UKRI, accessed February 16, 2023: https://gtr.ukri.org/projects?ref=710156

15. A. Bajpai, K. Rangra, D. Bansal, J. Electron. Mater. **50**, 7143 (2021)
16. N. Cennamo, F. Arcadio, and L. Zeni, in Proc. IEEE Sensors Applications Symposium (SAS) (IEEE, 2019).
17. J. Durisova, D. Pudis, M. Goraus, P. Gaso, Appl. Surf. Sci. **461**, 108 (2018)
18. J.L. Ninantay, A. Engler, J. Schwartz, P.A. Kohl, ECS J. Solid State Sci. Technol. **13**, 054004 (2024)
19. X. Wang, F. Yang, G. Guo, J. Li, Y. Sun, L. Zhang, Langmuir **40**, 11125 (2024)
20. K. Sano, Y. Ono, R. Tobinaga, Y. Imamura, Y. Hayashi, T. Yanagitani, A.C.S. Appl, Mater. Interfaces **16**, 22657 (2024)
21. D. N. Le, T. Park, S. M. Hwang, J. H. Kim, Y. C. Jung, N. Tiwale, A. Subramanian, W. Lee, R. Choi, M. M. Sung, C. Y. Nam, and J. Kim, Jpn. J. Appl. Phys. **62**, SG0812 (2023)

Chapter 2
Fundamentals of Photoresist Chemistry

2.1 Basic Chemistry of Photoresist Materials

Photoresists are light-sensitive materials used in lithography, and they function as imageable masks in etching, electroplating, and electroforming processes, thus enabling precise patterning in microfabrication [1]. It comprises a polymer (resin), a sensitizer (photoactive compound), and a solvent.

As shown in Fig. 2.1, resists are categorized based on resin composition as one-component (pure radiation-sensitive polymer) or multi-component (with additives). Based on exposure response, they are positive (exposed areas dissolve) or negative (exposed areas polymerize). Chemically amplified resists (CARs) use an acid-catalyzed reaction during post-exposure bake (PEB) to enhance solubility changes. Photoresist lithography remains dominant for high throughput, while EUV and e-beam lithography aim to achieve sub-10 nm resolution.

Characterizing photoresists involves evaluating their optical, chemical, mechanical, and lithographic properties to ensure optimal performance in patterning applications. Key parameters include resolution (smallest feature size), sensitivity (dose-to-clear for exposure efficiency), and contrast (sharpness of developed patterns) [2]. Line edge roughness (LER) and etch resistance maintain pattern fidelity and durability. Thermal stability and adhesion impact resist performance in post-processing. These characteristics, influenced by resin composition and photopolymerization, determine suitability for advanced lithography.

Photoresists function by undergoing chemical changes upon exposure to UV light, primarily altering their solubility in a developer solution. The fundamental process driving these solubility changes is photopolymerization, which occurs through radical or cationic initiation. The reaction causes significant alterations in solubility, adhesion, melting point, phase transitions, and both chemical and physical characteristics. For photopolymerization to occur, the system typically contains one or more reactive monomers (monofunctional or multifunctional), a photoinitiator (radical

K. Wang, *Photoresist Technology in Microsystems: Principles, Processes and Applications*, Springer Series in Advanced Microelectronics 15,
https://doi.org/10.1007/978-981-95-1606-3_2

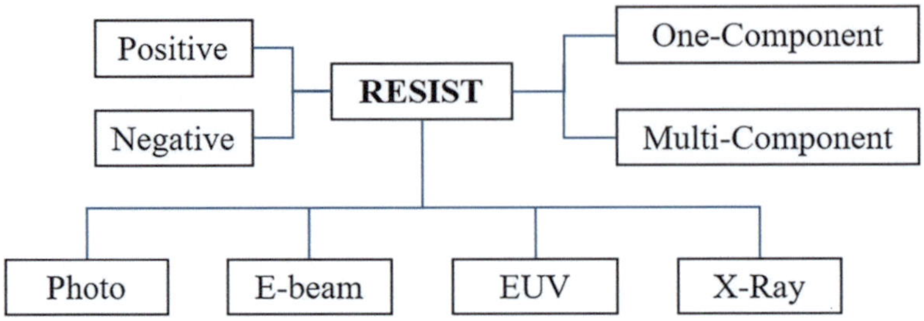

Fig. 2.1 Resist classification

or cationic), and optional components such as a prepolymer, stabilizer (to prevent premature polymerization), and solvent (depending on the application) [3].

RESINS: As the primary structural component of photoresists, resins determine mechanical stability, adhesion, solubility, and lithographic performance. They support photoactive compounds, ensuring adhesion to the substrate while maintaining chemical and thermal stability during processing. They comprises of monomers, which are typically monofunctional, bifunctional, or multifunctional acrylates and methacrylates [3]. Monofunctional monomers contain a single reactive site, allowing for linear polymer formation. Bifunctional monomers provide cross-linking, forming a three-dimensional polymer network, which enhances mechanical strength. Multifunctional monomers further increase cross-linking density, leading to higher hardness and reduced solubility in developers.

Resins must also exhibit good solubility in solvents to enable uniform film formation during spin-coating. The molecular weight and polarity of the resin influence its solubility, with lower molecular weight resins providing better uniformity, while higher molecular weight resins enhance mechanical strength. Additionally, solvent evaporation must be controlled to prevent surface roughness, which can degrade lithographic performance. The adhesion properties of the resin are critical, ensuring that the photoresist layer remains intact during development and subsequent etching or processing steps.

Different lithographic techniques require specific resin formulations. Novolac resins provide solubility control in i-line and g-line lithography. Polyhydroxystyrene (PHS) ensures high transparency in deep UV (DUV) resists, while methacrylate-based resins facilitate acid diffusion in chemically amplified resists (CARs) for DUV and EUV lithography. For e-beam lithography, polymethyl methacrylate (PMMA) and hydrogen silsesquioxane (HSQ) offer high resolution, whereas epoxy-based SU-8 resins are favored for MEMS applications due to their strong cross-linking [3, 4].

SENSITIZERS: The sensitizer enables the photochemical reaction in photoresists, functioning as a photoactive compound (PAC) in positive resists, a photoacid generator (PAG) in chemically amplified resists (CARs), or a photoinitiator in negative resists. In positive resists, PACs undergo photodecomposition, increasing solubility in the developer but not initiating polymerization. In CARs, PAGs release acid upon exposure, catalyzing polymer deprotection during post-exposure bake (PEB) but not directly polymerizing the resist [3]. Conversely, in negative resists, photoinitiators generate radicals or cations upon UV exposure, directly initiating polymer cross-linking and polymerization.

A good photoinitiator must effectively absorb UV light, generate radicals with high quantum efficiency, and exhibit thermal stability to prevent unintended polymerization. It should remain soluble in monomers and oligomers for uniform polymerization and must not cause discoloration (such as yellowing) during hardening. Additionally, low toxicity is essential for safety in practical applications. Typically, sensitizers contain one or more of the following compounds: benzophenones, thioxanthones, acetophenones, dicarbonyl compounds, and benzoin ethers.

Solvent Composition: The solvent plays a crucial role in photoresist formulation, influencing film uniformity, drying characteristics, and overall process performance. They dissolve the photoresist resin and photoactive compound (PAC) or photoacid generator (PAG), transforming the mixture into a liquid form suitable for spin coating onto a wafer, ensuring uniform film deposition. Additionally, solvents regulate film thickness by controlling viscosity, while their evaporation rate directly impacts coating uniformity and defect formation.

Solvents such as ethyl lactate, propylene glycol monomethyl ether acetate (PGMEA), and methylmethoxy propionate are widely used due to their commercial availability, safety, and balanced evaporation rates. Some solvents, such as 2-ethylhexyloxy propionate, function as cosolvents, enhancing the dissolution properties of the main solvent system [5]. Organic solvents such as xylene, cyclohexanone, chlorobenzene have good solvent properties but are not used today due to toxicological concerns.

After the application of photoresist by spin coating, the following steps are followed before etching or ion implantation: (1) Post-Apply Bake, (2) Exposure, (3) Post-Exposure Bake (for CARs), (4) Development, and (5) Postbake.

Post-Apply Bake (PAB): The PAB removes the excess solvent from the photoresist, which stabilizes the resist film. The following effects are observed: (1) The film thickness reduces, (2) development properties (dissolution rate, contrast) change (3) the adhesion is improved and (4) film is less susceptible to contamination [5]. As baking at high temperatures affect the PAC by decomposition and resin by crosslinking and oxidization, baking at an optimal temperature is essential.

Exposure: Upon UV exposure, photoresists undergo chemical transformations that alter their solubility in a developer solution, enabling precise pattern formation. In positive photoresists, such as diazonaphthoquinone (DNQ), the PAC is converted into a carboxylic acid which is soluble in a basic developer [4]. The resin experience polymer chain scission, increasing solubility, while negative resists undergo crosslinking, reducing solubility. In chemically activated resists, a photoacid catalyzes solubility changes during the post-exposure bake (PEB), enhancing sensitivity and resolution.

Development: During development, the resist undergoes selective dissolution based on exposure-induced chemical transformations. Positive resists rely on photodecomposition for solubility, negative resists undergo polymerization, and CARs leverage acid catalysis. The choice of developer chemistry and processing conditions significantly impacts pattern resolution, line edge roughness (LER), and feature stability. Aqueous bases are generally used as developers. Different strategies such as spin, spray and puddle development are employed.

Postbake: The postbake is used to harden the final resist image for further processes such as ion implantation or etching. The image becomes thermally stable by exposure to high temperature (120-150ºC) or high intensity deep-UV light that cross-links the resin polymer [5], removes any residual solvent, thereby increasing the adhesion to the substrate. This also prepares the wafer to be more vacuum compatible for other processes.

Practical Processing Insights: Optimal process conditions are crucial for reliable photoresist performance. Before coating, substrates are often dehydrated (baked at ~ 100 °C) and primed with an adhesion promoter such as HMDS to improve resist adhesion. Softbake (PAB) is typically performed at ~ 80–110 °C for a short duration (on a hotplate or in an oven) to remove most solvent without degrading the resist (from ~ 70% to ~ 15% solvent); insufficient baking can leave excess solvent (causing pattern collapse), whereas over-baking can induce premature crosslinking or photoactive compound decomposition. For chemically amplified resists, a post-exposure bake (PEB) around 150 °C is essential to drive the acid-catalyzed deprotection reaction to completion while minimizing acid diffusion blur [4]. A final hard bake (postbake) at higher temperature solidifies the developed resist patterns, enhancing etch resistance and adhesion for subsequent processing [5].

In addition to baking steps, clean processing practices greatly influence outcomes. A mild oxygen plasma "descum" is commonly used after development to remove any residual resist film in exposed areas, ensuring that open features are truly clear of debris. Such O_2 plasma treatments (or ashing) are also used after pattern transfer to strip the remaining resist. Prior to resist application, plasma cleaning can be applied to the wafer to eliminate organic contaminants and promote uniform resist coverage [6]. Maintaining a controlled environment (low humidity to prevent water uptake in the resist, and minimal ambient light to prevent unintentional exposure) further improves process consistency. These practical measures, combined with careful optimization of spin-coating, baking, and development conditions, help achieve reproducible high-fidelity lithographic patterns.

2.2 Positive Versus Negative Photoresist

Photoresists are fundamental materials during the microfabrication process. These chemicals are related to the photolithography process where pattern specific structures are made on the surface of a substrate especially the silicon wafer during the fabrication of semiconductors like integrated circuits (ICs), biosensors, and micro-electromechanical systems (MEMS). The photoresists are usually light sensitive materials composed of a sensitizer, a polymer and a solvent. When the silicon wafer is put under the UV light exposure, the polymer changes its structure, allowing the photoresist to form thin layers over the wafer surface [7]. As shown in Fig. 2.2, there are two types of photoresists classified as positive photoresists and negative photoresists.

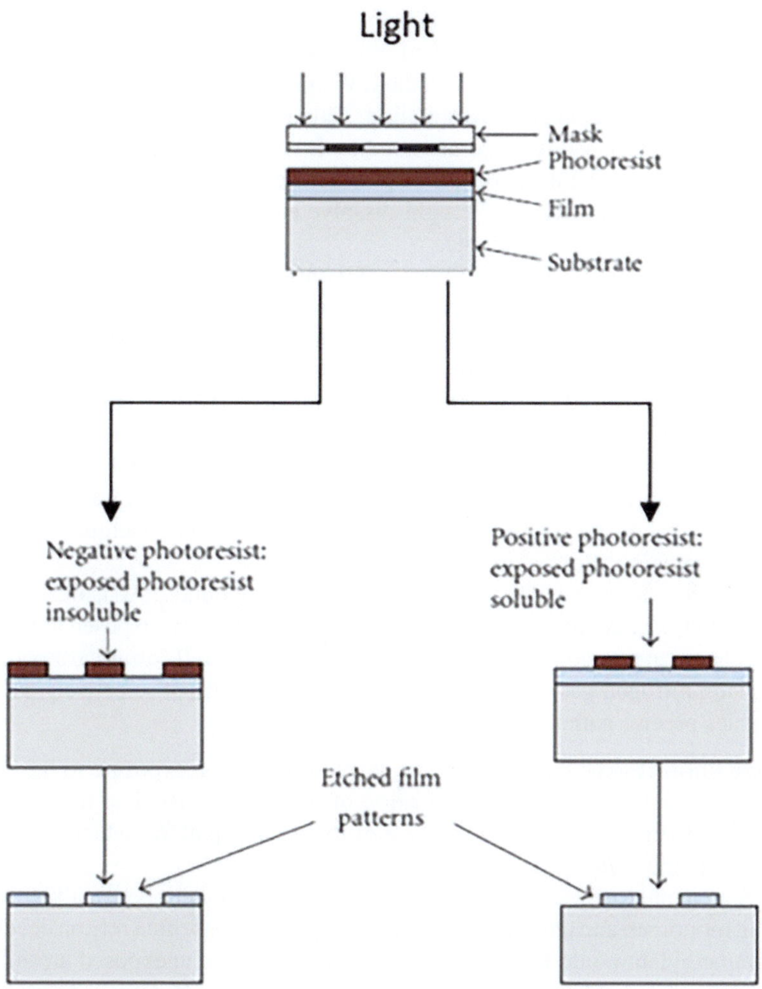

Fig. 2.2 Photolithography of positive and negative photoresist

Positive Photoresists: The most commonly used photoresists in IC manufacturing are positive photoresists due to high resolution and aqueous processing compatibility. Besides excellent resolution capability, there are also some key characteristics of these photoresists that make these resists ideal for manufacturing ICs like modern chemically amplified resists for sensitivity improvement and facile processing, including the use of aqueous developers [8].

In the positive photoresists, the polymer is weakened by exposure to UV light and the photochemical reaction occurs, making it more soluble in the developer. In this way, a positive pattern is achieved. In positive photoresists, the wafer has an exact pattern copy of the mask [7]. In this study [9], the detailed process of positive photoresists is clearly described. Positive photoresists consist of a resin and a photoactive compound (PAC). When the UV light is exposed over these chemicals, they go through a chemical transformation. This reaction converts the PAC into a soluble product. When PAC becomes soluble, it becomes easily dissolvable in the developer solution. After the exposure step, when the wafer is placed in the developer solution, the areas exposed to UV light dissolve easily in the developer, while the unexposed regions remain insoluble.

The resist is strongly attached to the wafer substrate, which makes it harder for the developer to remove the resist from the resist-substrate interface. As a result, it becomes harder for the developer to remove the resist at the bottom. On the other hand, resist interacts with air and undergoes surface inhibition which also slows dissolution at the top of the resist. Higher prebake (a heating step that removes solvent and stabilizes the resist) temperatures lead to premature decomposition of the PAC which plays a crucial role in resist behavior. This ultimately reduces dissolution rates. Overall, the development of positive photoresists is gone through by light-induced solubility changes which are controlled by some factors like reaction kinetics, developer diffusion and adhesion effects [9].

For example, as shown in Fig. 2.3, a photoactive compound (PAC), diazonaphthoquinone (DNQ), undergoes a chemical transformation after it is exposed to UV light. When a chemical reaction happens, DNQ reacts with water. It forms carboxylic acid by releasing nitrogen gas (N_2). In this manner, the photoresist selectively dissolves and enables precise pattern formation [10].

Negative Photoresists: In the case of negative photoresists, exposure to light, especially UV light, causes the polymerization of the photoresist. The mask used for negative photoresists contains the inverse of the desired pattern, which needs to be transferred to the wafer [7].

In negative photoresists, there are different chemical materials like alkali-soluble polymer, monomer, and photoinitiator. Negative photoresists often rely on developers like tetramethylammonium hydroxide (TMAH) to remove unexposed areas. When the photoresist is exposed to UV light, the photoinitiator triggers a cross-linking reaction. In this study [11], the role of TMAH in dissolving unexposed resist was discussed. The polymer is soluble in TMAH until it is exposed to UV light, but after exposure, the polymer becomes cross-linked and insoluble in TMAH. Then the cross-linked regions become insoluble. While the unexposed area of the wafer

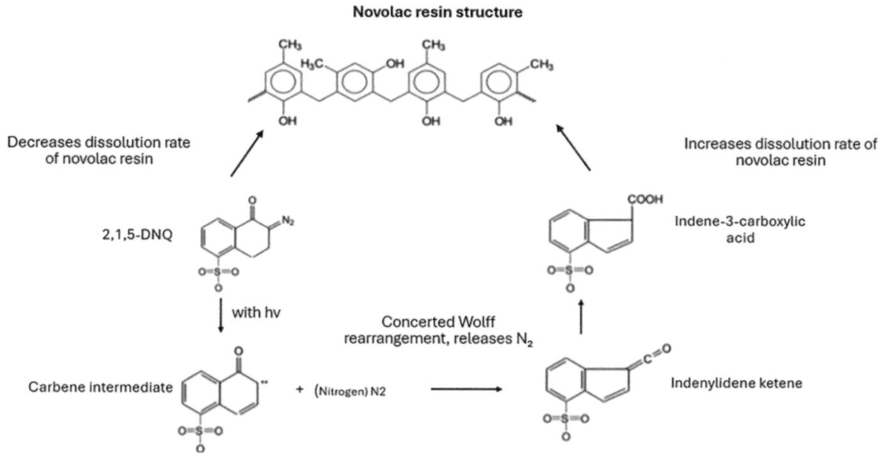

Fig. 2.3 UV-induced decomposition of PAC (DNQ) in positive photoresist

is in the development solution, the developer penetrates the resist film, creating a swelling layer that dissolves in a controlled manner. When resists are mixed with the developer, they gradually absorb the developer before breaking down into a soluble form.

Several factors influence dissolution. A higher acid value, which means an increased acid content in the polymer, makes the resist more soluble in the alkaline developer and speeds up dissolution. But the photoresists are more resistant to development at a lower acid value. Thicker films dissolve more slowly, and the addition of monomers accelerates development by weakening polymer interactions. This study highlights that dissolution in negative photoresists depends on cross-linking, and acid value [11].

Figure 2.4 illustrates the cross-linking reaction in a negative photoresist system, specifically involving cyclized polyisoprene, a common polymer used in negative photoresists. Upon UV exposure, a radical initiator (R-N) is generated, which subsequently reacts with the cyclized polyisoprene.

Three possible reactions that can occur are insertion, addition and abstraction in which insertion is the most common one. In insertion the radical inserts into the polymer chain, forming a new chemical bond. And then the radical attaches to the polymer, creating a more complex structure. A hydrogen atom is removed, generating a new radical site during abstraction. The resulting radicals lead to cross-linking, forming a network polymer. This cross-linking makes the exposed areas insoluble in the developer (e.g., TMAH), resulting the negative-tone resists remain on the wafer after development.

Applications:

Positive photoresists: Due to positive photoresist's ability to produce precise and intricate patterns, they are used in IC and MEMS manufacturing. Dry film positive

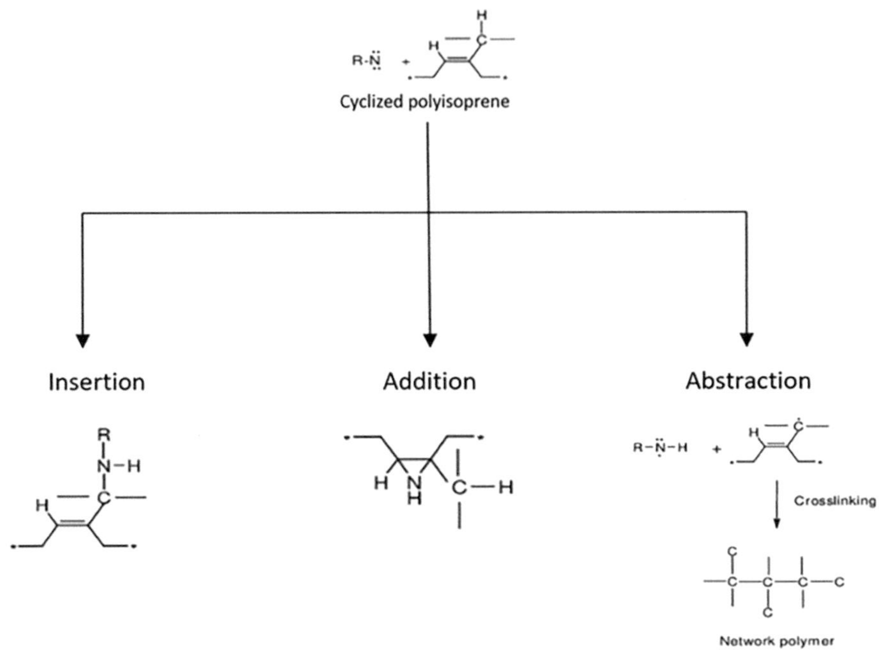

Fig. 2.4 Cross-Linking mechanism in negative photoresist development

photoresists are preferred in printed circuit board (PCB) manufacturing due to their ability to provide solvent-free processing, precise micropatterning, and improved adhesion to substrates. Positive photoresists are being developed to achieve unprecedented miniaturization, enabling feature sizes as small as 4 nm for advanced System-on-Chips. In addition to that, they are used for fabricating sub-10 nm semiconductor nodes since they provide the required resolution, sensitivity, and compatibility with the shorter 13.5 nm wavelength [12]. In addition to that, their ability to accurately transfer patterns onto substrates makes them suitable for creating complex optical components, such as grating couplers, waveguide structures such as multi-mode optical splitter and optical waveguide with surface Bragg grating in polydimethylsiloxane (PDMS) [13].

Negative photoresists: The SU-8 negative photoresist is widely used in various high-aspect-ratio (HAR) and three-dimensional (3D) microfabrication applications due to its excellent mechanical, thermal, and chemical stability, as well as its high-resolution patterning capabilities. It is also used in optical and photonic applications, including high-luminosity lighting panels and photonic crystals since it can be extended into the third dimension which is essential for manipulating light. In addition to that, HAR in biosensors help enhance sensitivity and throughput by increasing interaction sites for biomolecules, improving sensor performance. It is also used in Fabrication of micro- and nano-sized drug delivery structures due to its high mechanical and thermal stability which makes it an excellent material for drug delivery. Moreover, it

Table 2.1 Comparison of positive and negative photoresists: advantages & disadvantages [15, 16]

Factors	Positive photoresist [12]	Negative photoresist [14]
Resolution	Offers higher resolution with sharp, vertical sidewalks	Due to cross-linking discussion produces lower resolution, limiting fine-feature fabrication
Swelling & dimensional stability	No swelling during development, leading to better dimensional accuracy and stable resist profiles	May undergo swelling during development, affecting pattern uniformity and precision
Efficiency & processing	With controlled dissolution behavior, improved development times	This process needs longer exposure times and stronger development conditions, which decrease process speed
Cost	Because of complex processing and high-resolution requirements, this process is more expensive	More cost-effective with simpler processing steps and lower exposure energy requirements
Lithography compatibility	Compatible with EUV and DUV lithography, making it ideal for next-gen IC fabrication	Used in traditional photolithography, modern formulations provide better resolution control
Film thickness	It is more like limited to thin film applications	Applied in thicker layers
Etch & Chemical Resistance	Moderate etch resistance but may degrade under strong plasma etching conditions	Forms highly cross-linked structures, making it more resistant to acids, bases, and plasma etching

is also used to fabricate microfluidic devices for lab-on-a-chip applications, such as DNA analysis as it can lead to fabrication of precise microchannel structures [14]. For example, modern IC manufacturing exclusively employs chemically amplified positive resists (e.g., 193 nm and EUV CAR systems) to achieve the extremely small feature sizes and high throughput required for advanced chips. In contrast, MEMS fabrication often relies on thick negative resists like SU-8 to create high-aspect-ratio structures, valued for their mechanical robustness and stability [15, 16] (Table 2.1).

2.3 Advantages of Positive and Negative Photoresists:

2.4 Key Physical and Chemical Properties

Photoresists are specialized materials designed to respond to light exposure in photolithography, enabling precise pattern transfer onto semiconductor substrates. Their effectiveness is defined by several key physical and chemical properties, including resolution, adhesion, and etch resistance. These properties are critical for ensuring high-quality patterning and optimal performance in microfabrication. This section discusses these fundamental parameters, and the methods used to characterize them.

Resolution: In the context of photolithography and semiconductor manufacturing, resolution is a critical parameter that defines the capability to accurately reproduce fine patterns on a substrate. Resolution in photolithography refers to the smallest feature size that can be consistently and precisely patterned onto a semiconductor wafer. This encompasses the ability to delineate fine lines, spaces, and other intricate geometries essential for integrated circuits.

Resolution is a key factor in semiconductor manufacturing because it directly impacts several important aspects of chip performance and production. Higher resolution allows for the creation of smaller transistors, which results in faster and more energy-efficient chips. It also plays a crucial role in Moore's Law, as the continuous increase in transistor density depends on improvements in resolution. Additionally, better resolution enables feature miniaturization, allowing more components to fit into a smaller space, which enhances circuit density and overall functionality. Furthermore, high-resolution patterning reduces defects, leading to less material waste and greater fabrication efficiency, ultimately lowering costs and improving production yield [17].

Factors Affecting Resolution and Improvements: Resolution in optical lithography is influenced by various optical and process-related factors, primarily dictated by the Rayleigh criterion, which is expressed as:

$$CD = K_1 \frac{\lambda}{NA} \tag{2.1}$$

where CD is the critical dimension, λ is the exposure wavelength, NA is the numerical aperture, and k_1 is a process-dependent factor [18]. Reducing λ and increasing NA are primary strategies for enhancing resolution. Over time, lithography has evolved from G-line (436 nm) and I-line (365 nm) to KrF (248 nm), ArF (193 nm), and Extreme Ultraviolet (EUV, 13.5 nm) lithography, allowing for progressively smaller feature sizes [19]. However, factors such as diffraction limitations, resist performance, and mask complexity hinder further resolution improvements. Optical Proximity Correction (OPC) and Phase-Shifting Masks (PSM) have been widely adopted to mitigate these effects by enhancing image contrast and fidelity [18].

To overcome resolution limitations, several Resolution Enhancement Techniques (RETs) have been introduced. Double Patterning Lithography (DPL) and Computational Lithography have enabled sub-10 nm feature printing by splitting complex patterns into multiple exposures [18]. Off-Axis Illumination (OAI) modifies the incident angle of light to optimize diffraction patterns, improving line-edge resolution [18]. Additionally, advances in chemically amplified resists (CARs) have enhanced the sensitivity and pattern transfer capability of photoresists [17]. Emerging technologies such as High-NA EUV lithography and Machine Learning-based OPC promise further improvements in pattern fidelity and manufacturability [18].

Characterization methods for resolution in photolithography involve a combination of theoretical models and empirical measurements to evaluate the smallest feature sizes that can be reliably printed. Critical Dimension Scanning Electron Microscopy (CD-SEM) is widely used to measure line widths and detect variations in pattern fidelity. Atomic Force Microscopy (AFM) provides high-resolution surface profiling to assess feature dimensions and uniformity. Additionally, Modulation Transfer Function (MTF) analysis quantifies image contrast and sharpness, while Line-Edge Roughness (LER) and Line-Width Roughness (LWR) measurements determine the impact of patterning variations on device performance. Advanced optical simulations and lithography models also play a crucial role in predicting resolution limits and optimizing process parameters for next-generation semiconductor fabrication [20].

Adhesion: In photolithography and semiconductor manufacturing, adhesion refers to the ability of the photoresist layer to firmly attach to the substrate surface. Proper adhesion is crucial to ensure that the photoresist remains intact throughout various processing steps, such as development, etching, and implantation.

Strong adhesion is crucial in semiconductor manufacturing as it ensures pattern integrity, process reliability, and improved yield. It prevents the photoresist from peeling or lifting off the substrate, which is essential for maintaining the accuracy and quality of patterned features. Any detachment can introduce defects that compromise the final device's performance. Adhesion also plays a vital role in process reliability, as wafers go through various chemical and thermal processes during fabrication. Proper adhesion helps the photoresist withstand these conditions without degrading, ensuring a stable and efficient manufacturing process. Additionally, poor adhesion can lead to incomplete pattern transfer or feature distortion, increasing defect rates. By maintaining strong adhesion, manufacturers can achieve higher yields, minimize material waste, and reduce the need for costly rework or scrapping of defective wafers [4].

Factors Affecting Adhesion and Improvements: Several factors influence adhesion, including the chemical composition of the resist, substrate surface properties, and process conditions. Poor adhesion is often attributed to the inherent hydrophobic nature of photoresists, which struggle to bond with the typically hydrophilic silicon wafer surfaces. Additionally, contaminants, surface roughness, and the presence of native oxide layers can further reduce adhesion strength. To improve adhesion, surface treatment techniques such as dehydration baking and adhesion promoters like hexamethyldisilazane (HMDS) are widely used. These methods help modify the substrate surface, promoting stronger resist adhesion and reducing pattern collapse during development [4].

Process improvements in adhesion focus on enhancing the chemical interactions between the resist and the substrate. One key approach involves incorporating polar functional groups into photoresist formulations to improve their affinity for the substrate. Additionally, plasma treatments and UV ozone cleaning are used to modify the wafer surface energy, ensuring better adhesion. Advanced multi-layer resist systems with optimized bottom anti-reflective coatings (BARCs) also enhance adhesion by minimizing interface defects. Emerging research is exploring the use of nanostructured adhesion layers that create interlocking mechanisms at the resist-substrate interface, significantly improving adhesion strength in extreme ultraviolet (EUV) lithography applications [12].

Characterization of adhesion in photolithography is typically performed using contact angle measurements, which evaluates surface wettability and indicate the effectiveness of adhesion promoters like HMDS in improving resist-substrate interaction [4]. Additionally, tape tests (applying and removing adhesive tape over the resist to see if it peels) and scratch tests (cutting a grid into the resist and using tape to pull at it) are commonly used to assess adhesion strength by applying mechanical stress to the resist layer and observing delamination or peeling [12]. In addition, techniques like Fourier Transform Infrared Spectroscopy (FTIR) can verify the presence of adhesion promoter at the interface, and ellipsometry can monitor film thickness changes indicative of adhesion loss. These methods together ensure that adhesion is sufficient for subsequent processing steps.

Etch Resistance: Etch resistance refers to a photoresist's ability to withstand etching processes without significant degradation, ensuring the accurate transfer of patterns onto a substrate. In photolithography, after the photoresist pattern is developed, the substrate undergoes wet or dry etching to remove exposed areas while protecting regions covered by the resist. A highly etch-resistant photoresist must maintain its structural integrity under aggressive etchants such as halogen-based plasmas, acids, or ion bombardment. The chemical composition of the resist plays a crucial role in determining its etch resistance. Materials with aromatic structures and crosslinked polymers exhibit higher etch resistance compared to linear or aliphatic-based resists. [21].

Etch resistance is essential in semiconductor fabrication as it ensures pattern integrity, supports advanced lithography, controls surface roughness, and enhances device performance. It prevents dimension loss, deformation, and collapse during the etching process, maintaining the accuracy of patterned features. This is especially important for deep ultraviolet (DUV) and extreme ultraviolet (EUV) lithography, where thinner resists are required, making high etch resistance critical. Additionally, it helps control line edge roughness (LER) by minimizing depth variations and surface defects, ensuring precise patterning. Strong etch resistance also improves aspect ratio control, reducing substrate damage and contributing to higher production yields [21].

Factors Affecting Etch Resistance and Improvements: The etch resistance of a photoresist is influenced by chemical composition, structural properties, and process conditions. Aromatic ring-containing materials significantly enhance resistance by introducing stronger molecular bonds that withstand aggressive etching environments. One patent, EP 1 045 291 A2, by Schroeder, Kunkel, and Gutmann (2000), discusses how modifying resist composition with aromatic polycarboxylic acids, anhydrides, or acid chlorides improves etch resistance by creating a chemically reinforced polymer network. Additionally, crosslinking within the resist matrix enhances stability, preventing erosion under plasma etching conditions. The thickness of the resist layer also plays a role—thicker resists provide better protection but can impact resolution, necessitating a balance between resist thickness and etch resistance [21].

To improve etch resistance, various strategies are employed. The introduction of etch protectants containing aromatic functional groups increases chemical robustness while maintaining lithographic sensitivity. The use of plasma treatments to precondition the resist surface further enhances its ability to withstand etching without degradation. Process optimizations, such as post-exposure bake (PEB) and hard bake steps, improve the resist's thermal stability, further reducing etch erosion. Advanced resist formulations with multi-functional monomers create a denser polymer structure, increasing resistance to reactive ion etching (RIE) and halogen-based plasmas. As lithographic techniques progress toward sub-5 nm nodes, continuous improvements in photoresist chemistry and processing techniques will be essential to ensure pattern fidelity, etch selectivity, and overall fabrication efficiency [21].

Characterization methods for etch resistance involve measuring how well a resist withstands controlled etching processes. One approach is to perform etch rate tests, where a resist-coated wafer is subjected to a standard plasma etch (e.g., O_2 or SF_6, chlorine-based plasma) and the resist thickness loss is measured using ellipsometry or surface profilometry. A slower thickness loss indicates better etch resistance. End-point detection during plasma etch (monitoring optical emission or laser interferometry signals) can also reveal at what point the resist layer is breached. RIE simulations are sometimes used in research to predict how resist formulations will perform under different plasma chemistries. In addition, after etching, scanning electron microscopy (SEM) can inspect feature profiles to ensure that the resist has not been excessively eroded or distorted. These characterization techniques help in comparing resist materials and optimizing formulations for maximal etch durability [22].

2.5 Light-Induced Reactions: Principles and Mechanisms

Section 2.4 focuses on the fundamental photochemical events that drive structural changes within photoresist materials during lithography. Previously, in Sects. 2.1 through 2.3, we have examined the composition of photoresists, compared positive and negative tones, and analyzed the crucial physical and chemical properties. Building on that base, we now turn our attention to the core processes by which light initiates and propagates chemical transformations in the photoresist matrix. Understanding these mechanisms sets the stage for following chapters, which will concentrate on advanced processing steps, specialized formulations, and industrial-scale integration.

Photoresists owe their functionality to active compounds that absorb photons at specific wavelengths. When photons supply the appropriate energy, these compounds generate reactive species—commonly radicals or acids—that interact with the surrounding polymer or resin [23]. The details of these pathways hinge on both the nature of the sensitizer and the wavelength of illumination.

The photochemical chain of events begins when a photoactive compound (e.g., a photoacid generator or a radical initiator) absorbs a photon. Each photoactive species possesses a distinct absorption spectrum, typically matched to the exposure wavelength used in lithography systems. Absorption promotes the compound from a ground state to an excited state; this electronically excited species can then undergo various chemical reactions that do not occur under normal ground-state conditions [24]. In some cases, the excited species may dissipate heat or undergo unwanted fluorescence, but in photoresists, the key reaction pathways involve chemical transformations that generate reactive species [23, 25].

Two major categories of reactive species dominate traditional photoresist systems: free radicals and photoacid generators (PAG).

Free radicals play a key role in certain photoresist systems. In negative-tone resists, and some traditional positive resists, absorbing light of the right wavelength breaks chemical bonds in a photoinitiator. This process creates free radicals, which then cause either polymer cross-linking in negative-tone resists or the breaking of polymer chains in some positive-tone systems [26].

PAGs are commonly used in chemically amplified resists. When these molecules absorb light, they release a strong acid. This acid then triggers further chemical reactions, such as breaking or modifying polymer structures. Since a single acid molecule can drive multiple reactions, this process is called chemical amplification, which significantly increases the material's sensitivity to light [27].

In many modern photoresists, free-radical and acid-based mechanisms may be combined or further augmented by base generators, photosensitizers, or co-initiators that improve reaction efficiency [27]. Regardless of the precise mechanism, the physical outcome remains the same: selective solubility switching between exposed and unexposed areas.

Not every excited molecule forms a beneficial reactive species. Sometimes, oxygen present in the photoresist layer can quench radicals, reducing the overall efficiency of the photoreaction. Molecular oxygen can react rapidly with radicals or excited states, generating undesirable side products. In chemically amplified systems, acid diffusion can blur feature edges if it is not carefully controlled [28]. Hence, resist formulations often include additives or protective ambient conditions (e.g., inert gas environments) that minimize quenching by oxygen; well-controlled post-exposure bakes to manage acid diffusion [29].

Positive photoresists typically consist of a Novolac resin blended with a photoactive compound (PAC) which are hydrophobic and acts as a dissolution inhibitor, such as diazonaphthoquinone (DNQ). Upon exposure to ultraviolet (UV) light, the PAC absorbs photons and undergoes a structural transformation called Wolff rearrangement, that converts DNQ into a highly reactive ketene intermediate, which later hydrolyzes to produce carboxylic acid groups [30]. This reaction is the core photochemical pathway in positive resists, as the newly formed acid improves the polymer's dissolution in the alkaline developer [31]. This phenomenon is illustrated

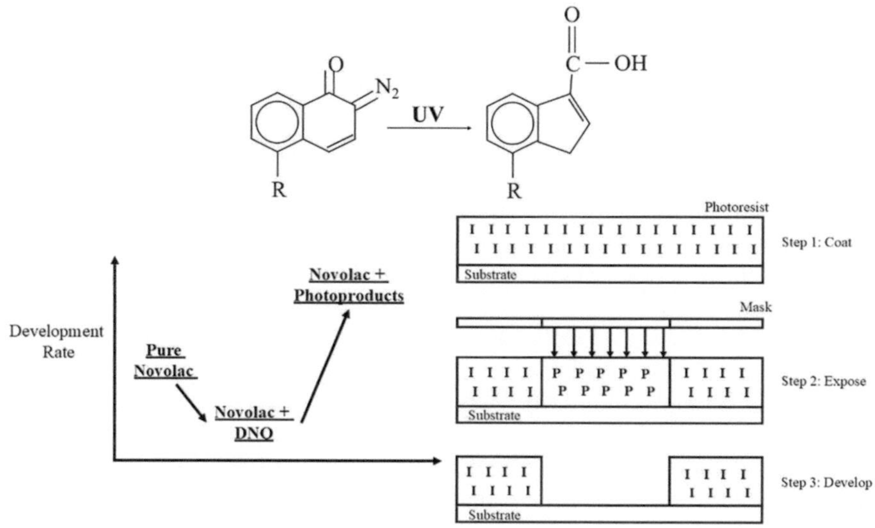

Fig. 2.5 Schematic picture of the mechanism by which DNQ-novolac resists function

in Fig. 2.5. As a result, the exposed regions become soluble and are subsequently washed away during development.

The main resin (e.g., Novolac) does not undergo major chemical changes itself; rather, it is the presence of the PAC's decomposition products that enhances the resin's solubility in the exposed areas. This synergy between the resin and the photoactive compound ensures that only those portions of the resist that have been exposed to sufficient UV intensity will dissolve. The uniformity and completeness of this conversion process are critical for achieving well-defined features on the substrate surface. Precise control over exposure energy and development conditions is crucial to maintaining consistent dissolution rates and pattern fidelity [32, 33].

In contrast, negative photoresists rely on photochemical pathways involving crosslinking. A common example is a resin (e.g., polyisoprene) containing sensitizers that generate reactive intermediates when exposed to UV light [32]. On a molecular level, in polyisoprene-based systems, the isoprene double bonds act as active sites for radical-mediated coupling, creating a three-dimensional polymer lattice. These intermediates initiate crosslinking reactions between the resin's polymer chains, causing the exposed areas to become increasingly insoluble in the developer. Because these reactions form a denser molecular network, the exposed regions remain on the substrate while the unexposed regions dissolve during development [33].

Key parameters influencing the extent of crosslinking include the intensity and duration of exposure, as well as the specific sensitizer chemistry. Small variations in these factors can significantly affect the degree of polymer networking, which in turn impacts the overall resist profile and thickness uniformity [34].

Chemically Amplified Resists: Reaction Mechanisms: Chemically amplified resists (CARs) leverage a catalytic chain reaction to induce solubility changes. Upon photon absorption, a photoacid generator (PAG) releases a strong acid (as described in Sect. 4.2.2). During a subsequent bake, this acid diffuses and repeatedly cleaves acid-labile protecting groups on the polymer resin [2]. A classic example is a polymer with tert-butoxycarbonyl (t-BOC) protecting groups: under acid catalysis, the t-BOC moiety is removed, yielding a polar phenolic hydroxyl group and volatile by-products (isobutylene and CO_2). This deprotection reaction drastically increases the resin's solubility in the aqueous developer for exposed regions. Chemical Equation:

$$\text{Ar} - \text{O} - \text{C(O)OC(CH}_3)_3 \rightarrow^{H^+} \text{Ar - OH} + \text{C}_4\text{H}_8 + \text{CO}_2$$

where 'Ar' is an aryl group [30]. Because each acid molecule can trigger many deprotection events before being quenched, CARs attain extremely high sensitivity via this amplification effect [2]. The trade-off is that acid diffusion must be carefully controlled to maintain feature sharpness and prevent line blur [4]. Table 2.2 summarizes the key reaction pathways for different resist types (DNQ-based positive, chemically amplified, and negative-tone) and their chemical outcomes.

Wavelength: As discussed earlier, the photoresist's performance and pattern fidelity are highly dependent on the wavelength of the light used for exposure. Common lithographic wavelengths include the i-line at 365 nm and deep ultraviolet (DUV) lines at 248 nm or 193 nm, each requiring photoresists engineered to absorb and respond optimally within these spectral ranges [32]. Generally, shorter wavelengths enable higher resolution because they can create smaller interference patterns and achieve finer feature sizes [35]. The relation between resolution and wavelength can

Table 2.2 Key photochemical reaction pathways in different resist types

Resist type	Photoactive compound/ initiator	Light-Induced reaction	Solubility outcome
Positive (DNQ-novolac)	Diazonaphthoquinone (PAC)	DNQ $\rightarrow$ ketene + N_2 (upon hν); ketene + H_2O $\rightarrow$ carboxylic acid [10]	Exposed areas become more soluble (PAC decomposition yields hydrophilic acids) [10]
Positive (CAR)	Photoacid Generator (e.g., onium salt)	PAG $\rightarrow$ acid (upon hν); acid + polymer-tBOC $\rightarrow$ polymer-OH + isobutylene + CO_2 [30]	Exposed areas become more soluble (acid-catalyzed deprotection increases polymer polarity) [30]
Negative	Photoinitiator + polymer (e.g., bisazide + polyisoprene)	Initiator $\rightarrow$ radicals (upon hν); radicals + C = C bonds $\rightarrow$ cross-linked network [14]	Exposed areas become insoluble (cross-linked polymer network resists developer) [14]

be observed in Eq. 2.1. However, shorter wavelengths can lead to increased photoresist degradation and stochastic effects, which may result in line edge roughness (LER) and line width roughness (LWR) [12]. In addition to that, shorter wavelengths often demand more sophisticated resist chemistries and optical systems to handle issues like absorption, scattering, and potential material degradation [32]. Photoresists must therefore be formulated accurately to absorb optimally at the chosen wavelength, balancing resolution requirements against process stability. Specifically, at 193 nm, many advanced resists can exhibit high absorption coefficients, limiting the effective penetration depth to less than a few hundred nanometers and requiring careful formulation to maintain consistent development profiles [32].

At the extreme ultraviolet (EUV) wavelength of 13.5 nm (used for next-generation lithography), photoresist behavior changes markedly [17]. EUV photons carry much higher energy, generating cascades of secondary electrons in the resist rather than direct photochemical transitions. As a result, stochastic effects become pronounced—for the same exposure dose, EUV provides an order of magnitude fewer photons than DUV, leading to significant photon shot noise in resist reactions. Moreover, most organic materials absorb EUV strongly, necessitating ultra-thin resist films (~20–50 nm) to avoid pattern blur and attenuation. While chemically amplified resists developed for 193 nm have been adapted for EUV, the combination of acid diffusion, higher absorption, and shot noise tends to increase line-edge roughness and random pattern defects at sub-20 nm features. To address these wavelength-dependent challenges, researchers are exploring new resist materials (including metal-containing hybrids to boost EUV absorption) and refining process conditions to improve sensitivity without sacrificing resolution or line-width roughness [4, 36].

Intensity: Light intensity, combined with exposure time, defines the exposure 'dose'. Higher intensity increases the rate of photon absorption and the generation of reactive intermediates (e.g., radicals or acid), which can lead to faster or more extensive polymer chain scission or crosslinking. Too little dose results in incomplete photochemical reactions, leaving portions of the resist unaltered and leading to poor pattern transfer. Conversely, excessive exposure can degrade or over-crosslink the resist, increase roughness (LER) which would reduce the resolution and complicate subsequent processing steps [32, 36]. This balance becomes particularly evident in the slope or sharpness of the pattern edges; a finely tuned dose ensures that the intended regions are precisely solubilized (for positive resists) or crosslinked (for negative resists), while limiting unwanted broadening or resist loss [32]. The ultraviolet exposure doses required for DNQ-novolac resists typically range from several hundred to several thousand millijoules per square centimeter (mJ/cm^2). This variation depends on factors such as the thickness of the resist layer, the specific wavelength of light used during exposure, and the resist's formulation [31].

Modern Advances in Photoresists: Molecular Glass Resists and Hybrid Materials: Continuing improvements in photoresist chemistry have yielded new classes of resist materials designed to meet the demands of ever-smaller feature sizes. One such innovation is the development of molecular glass resists, which are based on small, highly branched molecules that form amorphous films without long polymer chains.

Unlike conventional polymer resists, these molecular resists have a uniform molecular size (often < 2 nm) and lack chain entanglement, leading to reduced line-edge roughness and higher resolution potential [12]. Chemically amplified molecular glass resists have demonstrated sub-50 nm patterning with high contrast; for example, one study reported ~ 30 nm line/space patterns using a molecular glass formulation under EUV exposure. The absence of long polymer chains also mitigates internal stress and pattern collapse, although achieving adequate sensitivity and film robustness remains an active research area [37].

Another promising direction is the use of organic–inorganic hybrid resists, which incorporate metal or metal-oxide components into the resist matrix. These hybrids (for instance, blends of polymer matrices with nanoparticles or metal complexes) exploit the high EUV absorption and etch resistance of inorganic elements. By increasing the absorption cross-section at 13.5 nm, hybrid resists can improve sensitivity (fewer photons required) and simultaneously provide better resistance to plasma etching due to the inorganic content. Recent studies report hybrid resists based on materials like hafnium, zirconium, or aluminum oxides that achieve higher resolution with acceptable line-edge roughness at EUV exposures. Such resists can be engineered to remain positive-tone or negative-tone depending on whether the inorganic inclusion promotes solubility switching or crosslinking upon exposure. Both molecular glass and hybrid resist approaches are at the forefront of resist technology development, aiming to overcome the limitations of traditional polymer CAR systems in advanced lithography [38].

Future Challenges for Sub-10 nm Lithography: As lithography pushes to pattern features below 10 nm, photoresist chemistry faces profound challenges. At these dimensions, only a handful of polymer or molecular units define a single feature, which amplifies the impact of molecular-scale variations and statistical fluctuations (stochastic effects) on pattern fidelity [39]. One critical issue is the trade-off between resolution, line-edge roughness (LER), and sensitivity—often referred to as the RLS trade-off. Improving one aspect often degrades another: for instance, increasing resist sensitivity (by using more active compounds or higher acid diffusion) can worsen LER due to unwanted chemical spreading, while attempts to sharpen features can reduce sensitivity, requiring higher doses that risk damaging underlying layers [40]. Moreover, the scarcity of photons at EUV and beyond exacerbates shot noise, meaning some resist sites may not receive enough exposure to properly initiate chemical changes, leading to missing or deformed pattern elements [41]. To meet sub-10 nm requirements, ongoing research is exploring multiple strategies. High-NA EUV lithography will improve resolution but demands correspondingly advanced resist materials that can deliver precision at a tighter depth of focus and higher photon density. Novel resist platforms—such as the molecular glass and hybrid inorganic resists discussed above—are candidates to provide both higher resolution and better etch resistance than conventional CAR polymers. Additionally, post-lithography smoothing techniques (for example, vapor-phase diffusion of silane compounds to reinforce patterns) and multipatterning approaches can help overcome resist limitations at the cost of added process complexity. In the long term, achieving reliable patterning at sub-10 nm scales may require a combination of these material innovations and process techniques, as well as consideration of alternative patterning methods (such as directed self-assembly or nanoimprint lithography) if conventional photoresists reach their fundamental limits. The drive toward ever-smaller devices continue to make photoresist development a critical and challenging frontier in nanofabrication [42].

References

1. W. Guo, J. Chen, T. Ma, Z. Chen, M. Li, H. Zeng, J. Lu, Adv. Funct. Mater. **34**, 2310338 (2024)
2. W. Wei, L. Jingcheng, L. Hu, M. Qidao, L. Xiaoya, Progress in Chemistry **26**, 1867–1888 (2014)
3. M. Li, E. Aqad, Adv. Funct. Mater. n/a, 2420962 (2025)
4. X. Zhou, X. Liu, Z. Gu, Adv. Mater. **36**, 2409326 (2024)
5. V.M. Prabhu, S. Kang, D.L. VanderHart, S.K. Satija, E.K. Lin, W.-L. Wu, Adv. Mater. **23**, 388–408 (2011)
6. C. Sheng, X. Dong, Y. Zhu, X. Wang, X. Chen, Y. Xia, Z. Xu, P. Zhou, J. Wan, W. Bao, Adv. Funct. Mater. **33**, 2304778 (2023)
7. Y. Zhang, W.Z. Jiang, W. Jiang, X.Y. Zhang, J. Dong, Y.M. Xie, K.E. Evans, X. Ren, Adv. Funct. Mater. **35**, 2421746 (2025)
8. W.-M. Yeh, D.E. Noga, R.A. Lawson, L.M. Tolbert, C.L. Henderson, J Vac Sci & Technol B 28, C6S6-C6S11 (2010)
9. C.A. Mack, J. Electrochem. Soc. **134**, 148 (1987)
10. S.-H. Nam, G. Hyun, D. Cho, S. Han, G. Bae, H. Chen, K. Kim, Y. Ham, J. Park, S. Jeon, Nano Res. **14**, 2965–2980 (2021)
11. A. Tsuneishi, S. Uchiyama, T. Kozawa, Jpn. J. Appl. Phys. **57**, 046501 (2018)
12. M. Hassaan, U. Saleem, A. Singh, A.J. Haque, K. Wang, Recent advances in positive photoresists: mechanisms and fabrication, Materials, (2024)
13. J. Liu, Y. Yao, X. Li, Z. Zhang, Chem. Eng. J. **408**, 127262 (2021)
14. J.M. Shaw, J.D. Gelorme, N.C. LaBianca, W.E. Conley, S.J. Holmes, IBM J. Res. Dev. 41, 81–94 (1997)
15. L. Desmet, S. Van Overmeire, J. Van Erps, H. Ottevaere, C. Debaes, H. Thienpont, J. Micromech. Microeng. **17**, 81 (2007)
16. P. Kyoung-sun, K. Dae-yong, C. Sang-kuk, S. Dong Hack, A novel approach to chemically amplified resist materials for next generation of lithography, In: 2002 International Microprocesses and Nanotechnology Conference, 2002. Digest of Papers., p. 242 (2002)
17. D. Kazazis, J.G. Santaclara, J. van Schoot, I. Mochi, Y. Ekinci, Nat. Rev. Methods Primers **4**, 84 (2024)
18. M.S. Franklin, Resolution enhancement technology: the past, the present, and extensions for the future, Proc.SPIE, pp. 1–20 (2004)
19. Y. Tang, Highlights in Science. Eng. Technol. **29**, 61–68 (2023)
20. A.K. Wong, IEEE Micro **23**, 12–21 (2003)
21. S. Kishimura, Y. Sakai, K. Tsujita, Y. Matsui, Jpn. J. Appl. Phys. **38**, 250 (1999)
22. E. Gogolides, P. Argitis, E.A. Couladouros, V.P. Vidali, M. Vasilopoulou, G. Cordoyiannis, C.D. Diakoumakos, A. Tserepi, Journal of Vacuum Science & Technology B: Microelectronics and Nanometer Structures Processing. Measurement, and Phenomena **21**, 141–147 (2003)
23. X. Wang, P. Tao, Q. Wang, R. Zhao, T. Liu, Y. Hu, Z. Hu, Y. Wang, J. Wang, Y. Tang, H. Xu, X. He, Mater. Today **67**, 299–319 (2023)
24. S. O'Halloran, A. Pandit, A. Heise, A. Kellett, Adv. Sci. **10**, 2204072 (2023)
25. G. Lim, K. Lee, S. Choi, H.J. Yoon, Coord. Chem. Rev. **493**, 215307 (2023)

26. N. Corrigan, M. Ciftci, K. Jung, C. Boyer, Angew. Chem. Int. Ed. **60**, 1748–1781 (2021)
27. S. Chatani, C.J. Kloxin, C.N. Bowman, Polym. Chem. **5**, 2187–2201 (2014)
28. P. Yang, W. Yang, Chem. Rev. **113**, 5547–5594 (2013)
29. C. Decker, Polym. Int. **51**, 1141–1150 (2002)
30. S. Wang, Y. Song, L. Jiang, J. Photochem. Photobiol., C 8, 18–29 (2007)
31. L. Yeo, H.-C. Chang, W. Wen, 5–7 January 2009 Hong Kong 19, 175–176 (2009)
32. E.A. Costner, M.W. Lin, W.-L. Jen, C.G. Willson, Annu. Rev. Mater. Res. **39**, 155–180 (2009)
33. E. Sharma, R. Rathi, J. Misharwal, B. Sinhmar, S. Kumari, J. Dalal, A. Kumar, Nanomaterials **12**, 2754 (2022)
34. Z.-Y. Liu, Y.-Q. Shang, H.-C. Yu, H. Qi, Y. Zhang, J. Chen, Y.-L. Wu, Mod. Phys. Lett. B **34**, 2050319 (2020)
35. M. Rothschild, R.B. Goodman, M.A. Hartney, M.W. Horn, R.R. Kunz, J.H.C. Sedlacek, D.C. Shaver, Journal of Vacuum Science & Technology B: Microelectronics and Nanometer Structures Processing. Measurement, and Phenomena **10**, 2989–2996 (1992)
36. A.C. Fischer, F. Forsberg, M. Lapisa, S.J. Bleiker, G. Stemme, N. Roxhed, F. Niklaus, Microsyst. Nanoeng. **1**, 15005 (2015)
37. S.W. Chang, R. Ayothi, D. Bratton, D. Yang, N. Felix, H.B. Cao, H. Deng, C.K. Ober, J. Mater. Chem. **16**, 1470–1474 (2006)
38. C. Luo, C. Xu, L. Lv, H. Li, X. Huang, W. Liu, RSC Adv. **10**, 8385–8395 (2020)
39. H. Fukuda, J. Appl. Phys. 133, (2023)
40. J. Cen, Z. Deng, S. Liu, Polym. Chem. **15**, 4599–4614 (2024)
41. R. Brainard, P. Trefonas, J. Lammers, C. Cutler, J. Mackevich, A. Trefonas, S. Robertson. *SPIE*(2004)
42. K. Ronse, pp. 289–316 (2017)

Chapter 3
Microsystems Fabrication Processes

3.1 Photolithography: Core Technique in Microsystems Manufacturing

Photolithography is a fundamental fabrication technique that has matured in the fields of integrated circuits, semiconductors, and microelectronics manufacturing. Its widespread adoption exemplifies its crucial role in enabling precise and scalable production processes for microscale structures, including microfluidic and microelectromechanical systems (MEMS) [1–4]. While photolithography has proven essential, it is limited by its physical constraints, as it can only create planar, lateral, or two-dimensional patterns. This results in challenges when fabricating microscale structures with highly complex features. However, advanced photolithographic equipment, combined with appropriate etching techniques, has overcome these limitations, allowing the creation of intricate microsystems [3].

The primary objective of photolithography is patterning the substrate, typically composed of silicon, by using light. The light interacts with a previously applied layer of photosensitive material, transferring patterns from masks that either permit or block the light. A photomask, also called just mask, is a transparent plate, often made of quartz or soda-lime glass, with a patterned opaque layer of chromium or ferric oxide film [3, 4]. This opaque layer blocks light, transferring its pattern onto the photosensitive material. Mask sizes, standardized by the Semiconductor Equipment and Materials International (SEMI) organization [5, 6], typically range from 4 to 9 inches, depending on the specific application. A cost-effective option for prototyping microsystems using photolithography involves the use of plastic transparency films with printed opaque patterns [1]. In microsystem fabrication, multiple masks are employed, as it is difficult or impossible to fabricate complex topographical features in a single step with one mask. All masks used during the various stages of fabrication are collectively referred to as a "mask set" [5, 6].

In traditional photolithography, masks interact with ultraviolet (UV) light, where the transparent glass allows light to pass through, while the chromium film, typically

K. Wang, *Photoresist Technology in Microsystems: Principles, Processes and Applications*, Springer Series in Advanced Microelectronics 15,
https://doi.org/10.1007/978-981-95-1606-3_3

around 100 nm thick (note: different absorber film thicknesses may be used depending on the technique and requirements), absorbs the UV light. Additionally, masks are classified based on their “mask polarity,” which refers to whether the pattern of the absorber film material creates a masking island or hole. In an island pattern, the film material is surrounded by a clear field, creating an island that blocks light while allowing it to pass through the clear field. This is referred to as a light field mask. Conversely, in a hole pattern, the film material contains holes, allowing light to pass through the holes while blocking it in the surrounding areas. This is known as a dark field mask. Masks can also be oriented depending on which side the chromium film is deposited. When the chromium-coated side faces upward, the mask is referred to as having a Chrome-Up orientation, or Chrome-Down if the chromium-coated side faces downward. Chrome-Up and Chrome-Down orientations can be readily understood from Fig. 3.1. During photolithography, the chromium-coated side of the mask faces the photosensitive material and the substrate beneath it in a Chrome-Down orientation [1–4].

The photosensitive material exposed to UV light is referred to as photoresist. Similar to the mask, photoresist has a polarity, either positive or negative. The photoresist layer acts as an intermediary or sacrificial layer, chemically reacting to light exposure to create a mirror image of the mask pattern. In positive photoresist, the areas exposed to UV light degrade and can be washed away with a developer. In contrast, negative photoresist hardens when exposed to UV light, and the unexposed areas are removed by the developer. Determining the correct mask and photoresist polarity is crucial for successful photolithographic processes. Figure 3.2 illustrates the pattern transfer process and the importance of mask and resist polarity selection.

It is important to note that positive photoresists offer higher resolution due to reduced refraction effects, while negative photoresists provide greater resistance to wet etching and are cost-effective for prototyping microsystems. Although photoresists are typically sacrificial and used as intermediate materials, they are also employed as permanent materials in bio-microsystems and microfluidic devices [1]. For example, Roos et al. (2021) [2] describe the direct application of dry film photoresist over substrate material to create microfluidic channels as well as the application of the same photoresist material to modify microcantilevers.

Once the substrate has been cleaned to remove organic, inorganic, and metallic contaminants, the photoresist is evenly applied on top of it. The substrate is then soft-baked at a moderate temperature to remove solvent from the photoresist and improve adhesion. The subsequent photolithographic process can be summarized in

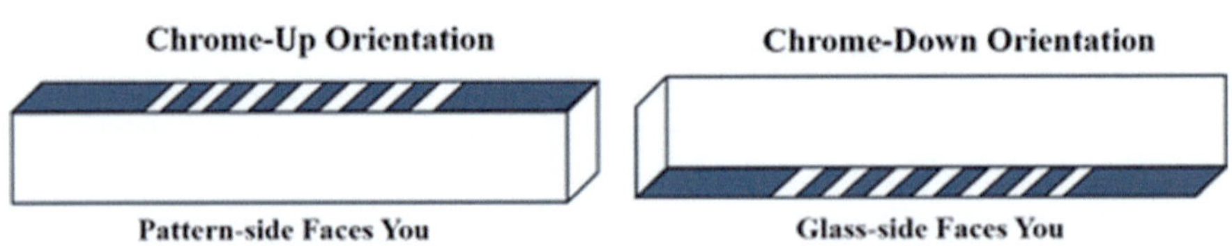

Fig. 3.1 Chrome-up and Chrome-down orientations

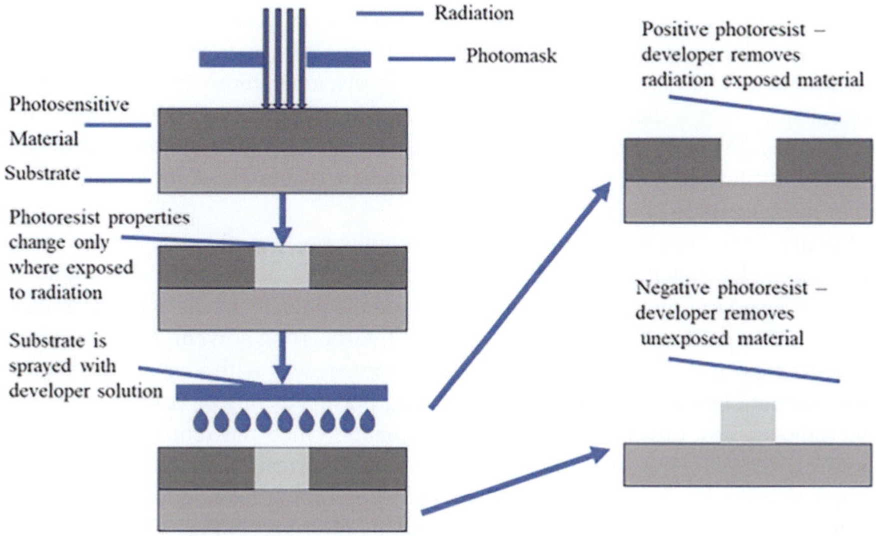

Fig. 3.2 Pattern transfer in positive and negative photoresists

three main steps. The process begins with the lateral positioning and alignment of the mask and the photoresist-coated substrate. The distance between the mask and the substrate must be carefully adjusted, with the gap varying depending on the specific photolithographic technique employed. In contact and proximity printing, the gap between the mask and the substrate is minimal, with no gap in contact printing. Both contact and proximity printing use a uniformly intense light source to illuminate the entire area of the mask and substrate simultaneously, enabling patterning of the entire surface in a single exposure. These techniques offer improved resolution due to the reduced or negligible gap between the mask and the substrate. The minimum resolvable feature size, r, can be determined by the following equation:

$$r = 1.5\sqrt{\lambda\left(s + \frac{z}{2}\right)}$$

where r is the minimum feature size, λ is the wavelength of the incident ultraviolet light, s is the distance between the mask and the photoresist surface, and z is the thickness of the photoresist layer. Contact and proximity printing are cost-effective for prototyping microsystems and provide high optical resolution. However, these methods are prone to damage on both the mask and the substrate, which limits their scalability for high-volume fabrication. In contrast, projection photolithography is employed for high-volume fabrication. This technique uses lenses to project a die-sized pattern from a reticle onto a specific location on the substrate [1–4].

The second step in the photolithography process is exposure, which has been extensively discussed. It is worth noting that different wavelengths of light can be employed, depending on the technique. For example, extreme ultraviolet (EUV) photolithography uses light with wavelengths ranging from 10 to 14 nm, while near-ultraviolet photolithography uses wavelengths between 350 and 500 nm. In near-ultraviolet photolithography, the g- and i-lines of a mercury lamp spectrum, corresponding to 435.8 nm and 365 nm, respectively, are commonly used. An important consideration is that shorter wavelength light results in reduced intensity, which necessitates the use of photoresists with higher sensitivity. While shorter wavelengths improve resolution, they also reduce the efficiency of the exposure system due to lower light intensity [1–4].

Finally, in the last step of the photolithographic process, the development stage takes place. During this step, the photoresist layer is developed using either an aqueous-based or organic solvent, depending on the polarity of the applied photoresist. Negative photoresist typically requires a hydrocarbon solvent, such as xylene, while the exposed positive photoresist can be washed away with a water-based developer. Further cleaning may be performed using plasma ashing or descum, a process that removes any unwanted photoresist remnants. Once the remaining patterned photoresist has served its purpose in other microfabrication processes, it can be etched or stripped away using either wet or dry stripping techniques. A high-level overview of the photolithographic process can be observed in Fig. 3.3.

In microsystem manufacturing, photolithography is often integrated with other fabrication processes to create the complex three-dimensional structures required for advanced microsystems, owing to its high precision, mass production capability, versatility, and cost efficiency. Researchers and students can also leverage photolithography for prototyping microsystems using the cost-effective options previously discussed [1–4].

3.2 Role of Photoresists in Patterning and Structuring

Photoresists, not to be confused with photoresistors, are photo-sensitive materials that change their chemical structure under exposure to light of a certain wavelength. Typically, a photoresist consists of three key components: a photoactive compound (the polymer), a base resin, and a suitable organic solvent system. Each plays a distinct role in the process. The polymer undergoes structural modifications when exposed to radiation, enabling pattern formation. The solvent ensures the photoresist spreads evenly across the substrate, creating a uniform thin layer. Meanwhile, the base resin gives the photoresist a certain amount of adhesion, sticking it to the surface of the substrate [7].

Photoresists serve as temporary protective materials that are used to define patterns on a substrate during photolithography with the help of a mask or used as sacrificial layers during the structural construction of a MEMs structure. In both cases, the photoresist usually does not form a part of the final MEMs structure, except for some exceptions [8]. The patterns formed with the help of the photoresists, when exposed to ultraviolet (UV) or deep ultraviolet (DUV) light, undergo chemical changes that determine their solubility in a developer solution. This property enables either material removal or deposition during patterning. The reaction of the photoresists to the

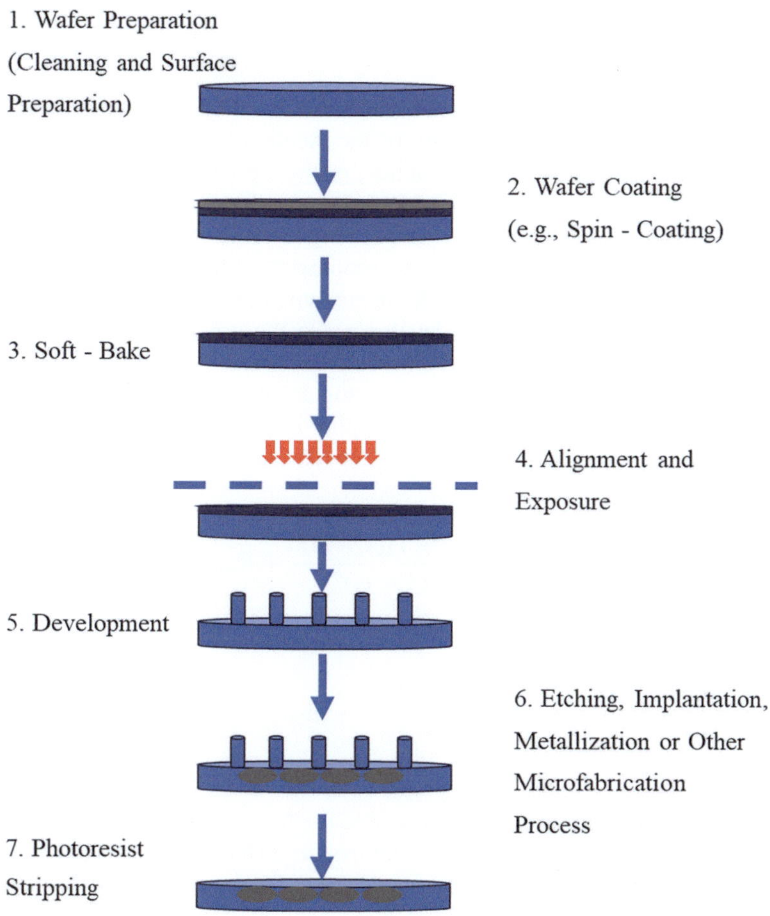

Fig. 3.3 A high-level overview of the photolithographic process

UV or DUV light defines the type of photoresist, i.e., positive or negative photoresist. A positive photoresist, when exposed to light, undergoes chain scission and becomes soluble in the developer solution. However, a negative photoresist, on exposure to light, hardens because of cross-linking, making it insoluble in the developer solution. In both cases, the soluble areas dissolve and leave behind a pattern on the photoresists, when the UV light passes through a mask on the photoresistive surface. A further explanation of the process is given in the next paragraph. Positive photoresists are known for their better adhesion leading to better and detailed patterning precision, but the negative photoresists are more robust and can support thicker films [9].

The sensitivity of photoresists to light follows the Dill Exposure Model [10]. This model explains how the absorption of light changes depending on the amount of exposure. The reaction follows the equation [10]:

$$\frac{\partial M(z,t)}{\partial t} = -I(z,t)M(z,t)C$$

where M(z,t) is a function describing the concentration of the photoactive compound in the photoresist, I(z,t) is the intensity of the incident light and C is an exposure parameter of the photoresist [9]. In practice, positive photoresists typically require an exposure dose in the range of 50–150 mJ/cm^2, whereas negative photoresists, due to their cross-linking nature, necessitate a higher dose of 200–400 mJ/cm^2. The polarity of the photoresist influences its response to light and plays a role in the lift-off process [9]. In positive resists, light exposure breaks down the PAC, making the exposed regions soluble in the developer. On the other hand, negative resists cross-link upon exposure, making them more resistant to dissolution. The model helps predict the behavior of photoresists during development, ensuring precise and controlled etching accuracy. To describe how photoresists develop after exposure, the Mack Model is used [11]. The development rate (R) can be mathematically expressed as:

$$R = \frac{R_{\max}(a+1)(1-m)^n}{a+(1-m)^n} + R_{\min}$$

and,

$$a = \frac{(n+1)}{(n-1)}(1-m_{TH})^n$$

where R_{max} is the maximum development rate, R_{min} is the minimum development rate, m_{TH} is the threshold concentration needed for development, and n represents the selectivity of the developer [11]. The development rate (R) of a photoresist depends on the concentration of the photoactive compound after exposure. It factors in parameters like maximum development rate, threshold concentration, and developer selectivity to predict how quickly the resist dissolves. This model is essential for controlling pattern formation and improving lithographic precision.

Positive photoresists contain diazonaphthoquinone (DNQ) compounds, which undergo photochemical decomposition when exposed to UV light. This process generates a ketene intermediate, which hydrolyzes into a carboxylic acid, making the exposed regions more soluble in an alkaline developer (e.g., Tetramethylammonium Hydroxide, TMAH) [12].

$$DNQ + h\upsilon \rightarrow Ketene\ Intermediate \rightarrow Carboxylic\ acid$$

$$Carboxylic\ acid + TMAH \rightarrow Soluble\ Compound$$

Negative photoresists, such as SU-8, utilize a photoacid generator (PAG) that, upon exposure to UV light, produces an acid catalyst. This acid initiates cross-linking polymerization during post-exposure baking (PEB), making the exposed regions insoluble in the developer [13].

$$SU - 8\,Monomer + h\upsilon \rightarrow Photoacid\ generation$$

$$photoacid + Post\ Exposure\ Bake \rightarrow Cross - Linked\ Polymer$$

Photoresists have been a game changer to allow complex stacking of different layers. During patterning and passivation, having sharp and defined edges (i.e. line-edge roughness (LER) of the photoresist pattern) ensures high-quality etching and better resolution. This is not only affected by the choice of photoresist but also by the choice of developer. Tetramethylammonium Hydroxide (TMAH) is widely used for positive photoresists, while organic solvents are common for negative photoresists. The chemical interaction between photoresists and developers affects line-edge roughness (LER) and resolution of the pattern [9]. In some cases, a combination of a thin top layer and a thicker bottom layer (dual-layer photoresists) may prove to be better for the edge roughness. Apart from this, it has been seen that EUV lithography, operating at 13.5 nm wavelength, allows for higher-resolution patterning (below 10 nm) compared to Deep Ultraviolet (DUV) lithography (193 nm) [18]. However, in this case, we may have to use chemically Amplified Resists (CARS) for their sensitivity and fast processing, possible with the help of polymer material engineering [14–16].

A thinner layer of the photoresist on the substrate can give out a higher resolution. This can be achieved by setting up particular spin levels and getting the desired level of thickness. During the exposure, diffraction effects sometimes result in poor resolution of the patterned edges. Also, Post-Exposure Baking (PEB) is one of the important ways to enhance the quality of the substrate and reduce the surface roughness for better resolution.

Modern MEMS devices and advanced semiconductor packaging often require multi-layer structures, demanding photoresists with precise adhesion and patterning properties. Adhesion and Passivation with the use of photoresists is an important step in the etching process. 3D MEMS ICs (Integrated Circuits) in current times are complex devices that require multi-layer lithography for which etching becomes a necessity [15, 16]. Photoresists have proven to be crucial for the structural integrity of the current MEMS devices hence playing an important role in the More-Than-Moore theory. Photoresists such as SU8 can also be used in the integration of CMOS circuits with MEMS sensors, through different wafer bonding techniques enabling varied applications[17].

3.3 Common Microsystems Fabrication Techniques

The term Bulk Micromachining represents all the micromachining processes that apply the concept of selective material removal from a bulk workpiece which is a substrate [18]. Like normal machining processes for regular sized workpieces, micromachining envelopes different techniques to remove material from a substrate. These techniques vary and diverge depending on the requirements, scale of production, and needs of the final product that is being prepared.

Bulk Micromachining Techniques [19]:

1. Wet Etching: In this process, wet etchants are used to remove the bulk material either dependent or independent on the material's crystal orientation.
   a. Isotropic etching: In isotropic etching, the liquid etches in all directions equally which results in round etching profiles. This results in bigger undercuts under the mask present on top of the substrate. The most unique characteristic about this type of etching is that it does not depend on the crystal planes' orientations of the substrate.
   b. Anisotropic etching: It can etch selectively based on crystal orientation. This usually results in V or U shaped etch profiles. Undercuts in this case are much less prominent than the previous case. Etching on (111) planes is much slower than it is on (100) and (110) which makes it suitable for being the plane parallel to the sidewalls.
2. Dry Etching: It replaces the liquid used in wet etching with gas to achieve different results.
   a. Reactive ion etching (RIE): uses gases that perform selected etching on the substrate's targeted area/s with high precision then an electric field is applied to the gas to excite it creates ionized gas (plasma). After that, the waste product is usually discarded as gas that is pumped out of the etching chamber.
   b. Deep reactive ion etching (DRIE) also known as Bosch process is an improved version of (RIE). It is well known for its capabilities of providing high aspect ratios up to 1:43 [20]. It works in an iterative sequence that consists of two steps, the first one is normal plasma etching where the substrate is exposed to ionized gas that reacts with it. After that, a passivation layer that stops ongoing etching is deposited onto the surface and protects the groove's sides from further etching resulting in wavy uniform sidewalls as seen if Fig. 3.4.
3. Laser Micromachining: As its name suggests, it uses focused laser beams to perform the etching process. The most recognized feature of this method is non-contact machining and the ability to work on different materials not just silicon. It provides very high precision micro to sub-micron resolution [21] (Table 3.1).

Common Challenges

- In DRIE, one of the most common challenges during the process is (RIE lag) or aspect-ratio-dependent etching (ARDE) [21]. ARDE is a phenomenon where

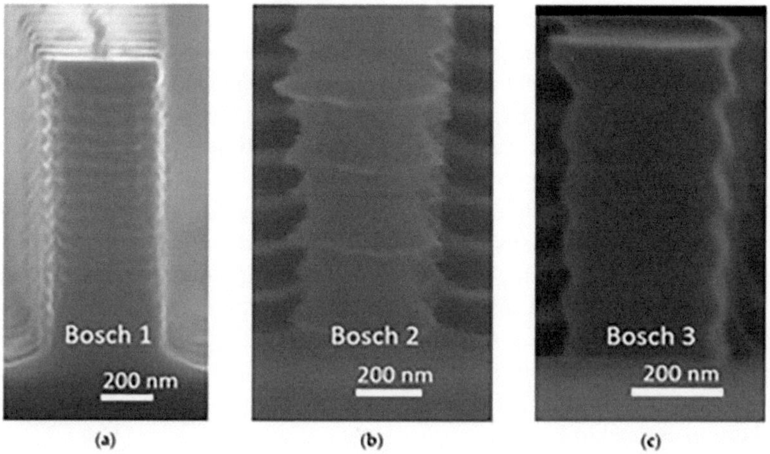

Fig. 3.4 DRIE Sidewalls [20]

Table 3.1 Comparison between different bulk micromachining techniques based on previous discussion

Feature	Wet etching	Dry etching	Laser micromachining
Etching medium	Liquid etchants	Plasma (ionized gas)	Focused laser beams
Material removal mechanism	Chemical reaction dissolves material	Ion bombardment & chemical gas reaction with substrate	Material vaporization
Types	Isotropic: Uniform etching in all directions - Anisotropic: Selective etching based on crystal plane direction	- RIE: Precision plasma etching - DRIE (Bosch process): High-aspect-ratio etching with alternating etch/ passivation steps	Short-pulse lasers: Non-contact precise beam
Etch directionality	- Isotropic: No direction control - Anisotropic: Crystal-plane selective	- Isotropic: BrF_3, XeF_2 - Anisotropic: F-based plasmas	Can be controlled by laser movement
Aspect ratio control	Depends on crystal planes	High (up to 43:1 in DRIE)	Depends on laser focus and intensity
Undercut formation	-High in isotropic etching - Low in anisotropic etching	Minimal in RIE, some in DRIE due to footing	None
Material compatibility	Mostly silicon (Si), oxides (SiO_2), and some metals	Si, oxides, nitrides, polymers, and some metals	Works with a wide range

etch rate depends on aspect ratio of the trench. It happens mainly due to Ion Flux Reduction which means that a lower number of ions reach the bottom of the trench due to its high aspect ratio which slows down the etching process. Another main reason for ARDE is byproduct accumulation at the bottom of deep trenches. These byproducts cannot escape easily from the bottom which partially blocks the etchant and reduces etch rate [22].

- Another common challenge present in dry etching is footing, also known as notching. Notching happens in SOIs where an insulator layer is underneath the etched silicon layer. When positive ions that are originating from the process gas hit the substrate material (usually silicon) which is a conductive material, they etch away that conductive material. However, when they hit the insulator material present in SOI, plasma ions stay trapped on the surface which deflects and repels the ions coming after [23]. These deflected ion trajectories cause undercutting at the bottom as seen in Fig. 3.5.

Surface Micromachining refers to the technique of building microstructures layer by layer using patterning and deposition on top of a substrate. Layering is done by adding a sacrificial temporary and removable layer, then depositing the structural permanent layer that forms the final structure. Deposition methods [24]:

- Physical Vapor Deposition (PVD): PVD is done by converting the material desired for deposition into a vapor form by thermal evaporation or sputtering. Then, the vapor is condensed on top of the substrate to form a thin layer.
- Chemical Vapor Deposition (CVD): This deposition technique works by heating the substrate and exposing it to the gas that contains the desired material for the film. After that, the gas reacts with the substrate's surface and forms a film. CVD provides very controlled step coverage which makes it suitable for producing high aspect ratio layers

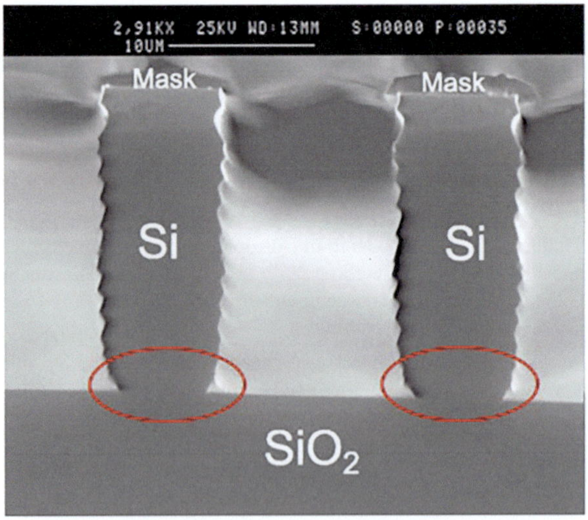

Fig. 3.5 Footing effect [22]

- Oxidation: Another form of layer deposition is the process of oxidizing the wafer substrate by applying high temperature along with H_2O or O_2 to grow an oxide film that can be used later as a sacrificial layer or a protection and passivation layer.

Patterning can be done after depositing each layer depending on the application using processes similar to the ones that are used in bulk micromachining. Photolithography defines the desired patterns to be removed or kept while etching removes the excess material to form the needed shape. The last step of each iteration is removing the sacrificial layer to produce the final form of the microstructure. The removal process is dependent on the sacrificial layer itself as the used etchant will vary for each material.

After classifying the two main micromachining techniques, we have a general overview of what each one of them represents and how they differ in forming structures.

Feature	Bulk micromachining	Surface micromachining
Fabrication Method	Selective removal of material from the substrate	Layer-by-layer deposition and patterning
Material Usage	Substrate (usually silicon) is machined	Thin films are deposited on a substrate
Process Techniques	Wet/dry etching, laser machining	CVD, PVD, oxidation
Aspect Ratio	High aspect ratios providing deep trenches, and cavities	Aspect ratio is based and variable on the thin-film itself
Undercuts	Present in isotropic etching	Minimal to none
Applications	Both techniques are usually incorporated during the microfabrication of the same product based on the requirements of the final result. Some of the applications include pressure sensors, accelerometers, microfluidics, high-precision structures, micro-mirrors, micro-switches, and IC fabrication	

LIGA is an acronym that originates from the German words Lithographie, Galvanoformung, and Abformung. This process is well renowned for its ability to provide high aspect ratio structures while maintaining high precision and smoothness of sidewalls [25].

LIGA Steps [25]:

- Photoresist Application: The very first step of LIGA is applying the photoresist material on the substrate. The photoresist usually used for LIGA is PMMA (Polymethyl Methacrylate) because of its ability to achieve nano patterning resolution and sharp walls.
- Irradiation (X-Ray Exposure): The mask is placed on top of the PMMA layer and then exposed to synchrotron-sourced high energy X-Rays for an extended time to affect the whole height of the thick photoresist layer.
- Development: A developer is applied to the photoresist layer to remove the exposed area.

- Electroforming: A metal undergoes electroplating into the developed and removed volume and on top of the substrate itself.
- Photoresist removal: The photoresist layer is removed. Most commonly for PMMA, it is removed using chemical dissolution.
- Application dependent: After acquiring the metal parts, it can be either used directly for its application, or it can be used as a mold for mass production processes like embossing or molding.
- Direct use: It can be used directly in different applications like microgears or high-frequency switches.
- Mold use: It can be used as a mold for plastic injection molding where the mold is filled with plastic material then separated from the metal body as seen in steps 5 and 6 in Fig. 3.6.

LIGA Applications:

1. RF MEMS Devices: Deep narrow trenches that are crucial for RF MEMS Devices can be microfabricated using LIGA [25].
2. Variable Capacitors: MEMS based capacitors that are tunable can be fabricated using LIGA process [25].
3. Biomedical Microstructures and Microfluidics: Can be used in construction of microchannels and pathways to deliver fluids. Can also be used for DNA sequencing and lab-on-chip systems [26].
4. Micro-optics and Photonics: LIGA can be used to create micro-lenses and waveguides for optical applications [27].
5. Polymer Injection Molding with Microstructures: Can be utilized for the production of microcantilevers [27].

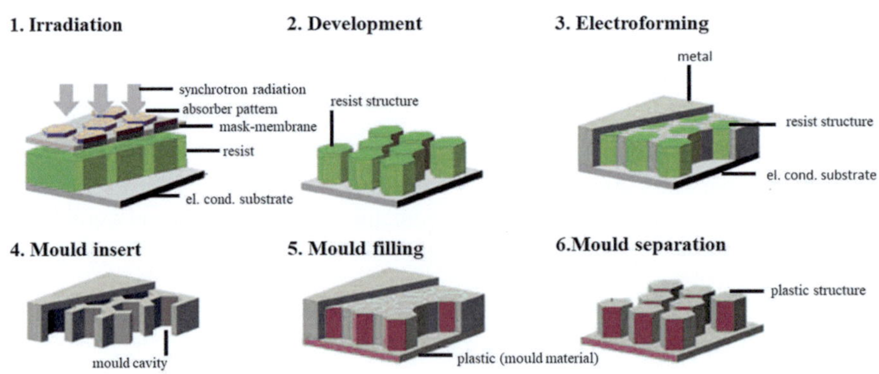

Fig. 3.6 Pattern transfer in positive and negative photoresists

3.4 Integration of Photoresists with Other Processing Steps

Photoresists play a critical role in various microsystem fabrication processes, including etching, deposition, and bonding. Each of these processes requires specific types of photoresists, which are utilized in distinct ways to achieve precise pattern transfer and material modifications.

There are two types of etching in microfabrication, dry etching and wet etching. For dry etching, a material is eroded by using the atoms or molecules of a partially ionized gas. To define a desired photoresist mask, the previous photoresist process has some specific requirements. The vertical sidewalls of the photoresist need to have the steepest photoresist profiles, which requires a high-contrast photoresist, as well as optimized process parameters for this high contrast. The AZ® 701 MiR photoresist is ideal for films with thickness below 1 μm, the AZ® ECI 3000 series is better for a thickness between 1 and 3 μm, and the photoresist AZ® 9260 is used for even thicker layers [28]. It is also advised to keep a high development rate for the photoresist and to use positive photoresist. It is also recommended to use a highly diluted developer, such as the AZ® 326 and AZ® 726 or the AZ® 400 K, or AZ® 351B. If the shape of the substrate needs to be an ellipsoid, the rectangular cross-section of a photoresist is softened by heating it up until reaching the softening temperature of the photoresist, which is approximately 100 - 110 °C for the positive photoresists AZ® 1500, AZ® 4500 and 9200. After dry etching, all standard strippers are suitable for removing the photoresist that is left on the substrate.

In the process of wet etching, there are different specifications for the photoresist. Firstly, the photoresist needs to have an optimum adhesion to the substrate to minimize lateral under-etching. For thicknesses between 500 nm and 3 μm, the most suitable photoresist would be the AZ® 1500 series; for thicknesses between 1–4 μm the ideal photoresist would be the AZ® ECI 3000 series; and, for thicknesses up to 10 μm, the ideal resist would be the AZ® 4500 series. For a high adhesion but with low demands on resolution, the PL 177 photoresist is the best option. Another factor to consider is when the etching solution contains hydrofluoric acid, because the fluoride ions can diffuse through the photoresist mask to the substrate. In this case, it is advised to use resists that are thick enough, such as the AZ® 4562, the AZ® 9260 or the AZ® 40 XT[29].

For deposition, other factors need to be considered in the photolithography process. First, it is advised to use a negative profile (negative photoresist and mask) to create a discontinuous layer of metal on top of the photoresist, which is required for the metal lift off [30]. In electroplating, the process that uses an electric current for thin metal deposition, there are basic requirements for the photoresist. First, the photoresist needs to be chemically stable. The cresol resin in AZ® and TI positive photoresists protects the photoresist layer in acidic solutions, but when the PH gets higher than 10, the photoresist can be affected. To reach alkaline stability, the photoresist mask can be hard baked above 140 °C, with the downside that the photoresist profile may be affected by the high temperatures. Another solution is cross-linking (exposing the photoresist to UV light) negative photoresists instead of positive [31].

Also, the photoresist needs to have a strong adhesion to the substrate to prevent peeling and underplating. To achieve this, some suitable photoresists would be AZ® 1500, 4500, 9200 or ECI 3000. Also, negative photoresists like AZ® nLOF 2000, AZ® 15 nXT or AZ® 125 nXT can be used. It is also important to adjust the soft-bake and hard-bake parameters to get proper adhesion. For soft-bake, a proper temperature would be of 100 °C for one minute/μm of photoresist film thickness on a hotplate. For hard-bake, the advice temperature would be between 120 °C and 140 °C, nevertheless, this can result in damage to positive photoresists. Another solution for improving adhesion is adding a thin coating of titanium or chromium in the seed layer, this additional layer needs to be removed after the development of the photoresist [32] (Fig. 3.7).

After soft-baking, in typical conditions, 2%-4% of solvent can be found in the photoresist film, this can affect metal deposition rates if the solvent is released into the electrolyte. To prevent this from happening, a prolonged or with higher temperatures soft-bake is recommended. Increasing the hard bake temperature for negative photoresists is also an option, but when the temperature is too high, since negative photoresists do not soften, this could make the photoresist less removable after electroplating.

Fig. 3.7 Silicon wafer coated with photoresist on a precision hot plate

There is a strong relationship between wafer bonding and photoresist application, especially in the MEMS context. Photoresists are often used as adhesive materials for wafer bonding, particularly negative photoresists such as SU-8. To bond two wafers together, a layer of photoresist is placed between them, this layer is later cured, creating a strong adhesive bond [33].

Photoresists are also used to create patterns that define areas where bonding should occur or should be avoided, this step is crucial for specific MEMS structures. Photoresists are also used for thermal and electrical insulation. Finally, when photoresists are used for bonding, lower bonding temperatures can be used, which allows the use of different materials with lower melting temperatures, making it suitable for different applications [33, 34].

The following table displays a summary of the type of photoresist recommended for each process step, the specific photoresists that should be used, and the process details.

Process	Photoresist type	Process details and specific photoresists
Dry etching	Positive	- High-contrast photoresists required for different thickness: < 1 μm (AZ® 701 MiR), 1–3 μm (AZ® ECI 3000), > 3 μm (AZ® 9260) - High development rate required - Highly diluted developer: AZ® 326, AZ® 726, AZ® 400 K, AZ® 351B - Resists needed for a softening temperature between 100 and 110 °C: AZ® 1500, AZ® 4500, AZ® 9200 - Any standard strippers can be used for removal
Wet etching	Positive	- High-contrast photoresists required for different thickness: 500 nm-1 μm (AZ® 1500 series), 1–4 μm (AZ® ECI 3000), 4 μm-10 μm (AZ® 4500 series) -Photoresist used for high adhesion and no high resolution required: PL 177 - with solutions contain hydrofluoric acid, a thick photoresist is required (AZ® 4562, the AZ® 9260 or the AZ® 40 XT)
Deposition	Negative	- Chemically stable photoresist - Hard bake above 140 °C for alkaline stability in solutions above a PH of 10 - Strong adhesion to substrate - Soft-bake: 100 °C for 1 min/μm - Hard-bake: 120–140 °C - Thin coating of titanium or chromium for adhesion improvement
Wafer bonding	Negative	- Used as adhesive material - Layer of photoresist placed between wafers - Cured to create strong adhesive bond - Defines bonding areas - Provides thermal and electrical insulation - Allows lower bonding temperatures for different materials

Certain aspects need to be adjusted on the photoresist in some specific scenarios during the processes of dry etching, deposition, and bonding. For dry etching, when extreme temperatures are used, the removability of the photoresist mask can be affected. Using high performance strippers like TechniStrip® P1316 for positive photoresists or the TechniStrip® NI555 for negative photoresists can help when standard removers are not enough. Heat development during etching can cause the edges of the photoresist to soften, some measures that can be taken to prevent this are: a sufficiently high heat buffer, reduced deposition rates with cooling intervals in between, and sufficient soft-bake on the photoresist to guarantee that most of the solvent evaporates, among others.

References

1. Z. Chen, Y. Li, Z. Man, A. Tang, Nano Res. **17**, 10386–10411 (2024)
2. M.M. Roos, A. Winkler, M. Nilsen, S.B. Menzel, S. Strehle, Int. J. Precis. Eng. Manuf.-Green Technol. **9**, 43–57 (2022)
3. W. Guo, J. Chen, T. Ma, Z. Chen, M. Li, H. Zeng, J. Lu, Adv. Funct. Mater. **34**, 2310338 (2024)
4. M. Esashi, Microsyst. Nanoeng. **7**, 60 (2021)
5. S.-J. Jeong, J.Y. Kim, B.H. Kim, H.-S. Moon, S.O. Kim, Mater. Today **16**, 468–476 (2013)
6. D. Kazazis, J.G. Santaclara, J. van Schoot, I. Mochi, Y. Ekinci, Nat. Rev. Methods Primers **4**, 84 (2024)
7. D.N. Le, T. Park, S.M. Hwang, J.-H. Kim, Y.C. Jung, N. Tiwale, A. Subramanian, W.-I. Lee, R. Choi, M.M. Sung, C.-Y. Nam, J. Kim, *Jpn. J. Appl. Phys.* 62, SG0812 (2023)
8. P.D. Cabral, T. Domingues, G. Machado, A. Chicharo, F. Cerqueira, E. Fernandes, E. Athayde, P. Alpuim, J. Borme, Materials **13**, 5728 (2020)
9. M. Hassaan, U. Saleem, A. Singh, A.J. Haque, K. Wang, Recent advances in positive photoresists: mechanisms and fabrication, Materials, (2024)
10. F.H. Dill, W.P. Hornberger, P.S. Hauge, J.M. Shaw, Electron Devices IEEE Trans On **22**, 445–452 (1975)
11. C.A. Mack, G. Arthur, Electrochem. Solid-State Lett. **1**, 86 (1998)
12. C.A. Mack, J. Electrochem. Soc. **134**, 148 (1987)
13. S. Chatani, C.J. Kloxin, C.N. Bowman, Polym. Chem. **5**, 2187–2201 (2014)
14. T. Manouras, P. Argitis, Nanomater. **10**, 1593 (2020)
15. I.S. Garcia, C. Ferreira, J.D. Santos, M. Martins, R.A. Dias, D.E. Aguiam, J. Cabral, J. Gaspar, J. Microelectromech. Syst. **29**, 734–740 (2020)
16. H. Li, Microsyst. Technol. **22**, 223–237 (2016)
17. Y. Ma, S. Wang, Z.-S. Wu, Nano-Micro Lett. **17**, 105 (2025)
18. R. Oboe, R. Antonello, *IntechOpen*, Rijeka, (2011)
19. M.S. Gerlt, N.F. Läubli, M. Manser, B.J. Nelson, J. Dual, *Micromachines* 12, (2021)
20. A.M. Baracu, C.A. Dirdal, A.M. Avram, A. Dinescu, R. Muller, G.U. Jensen, P.C.V. Thrane, H. Angelskår, Micromachines **12**, 501 (2021)
21. A. Gujrathi, S. Dongre, H. Nandan, *International Journal of Engineering Science Advanced Computing and Bio-Technology* (2021)
22. V. Ishchuk, B.E. Volland, M. Hauguth, M. Cooke, I.W. Rangelow, *J. Appl. Phys.* 112, (2012)
23. F. Laermer, A. Urban, Plasma Processes Polym. **16**, 1800207 (2019)
24. M. Huff, *Micromachines (Basel)* 13, (2022)
25. E.W. Becker, W. Ehrfeld, P. Hagmann, A. Maner, D. Münchmeyer, Microelectron. Eng. **4**, 35–56 (1986)
26. W. Luo, H. Han, R. Yu, L. Cai, R. Gao, Appl. Therm. Eng. **212**, 118558 (2022)
27. M. Bohling, T. Seiler, B. Wdowiak, J. Jahns, J. Mohr, M. Börner, Appl. Opt. **51**, 5989–5995 (2012)

28. Y. Zhang, Z. Hou, C. Si, G. Han, Y. Zhao, X. Lu, J. Liu, J. Ning, F. Yang, Micromachines **14**, 306 (2023)
29. B. Chang, P. Leussink, F. Jensen, J. Hübner, H. Jansen, Microelectron. Eng. **191**, 77–83 (2018)
30. D. Berkoh, S. Kulkarni, IEEE Trans. Semicond. Manuf. **32**, 513–517 (2019)
31. D. Hii, D. Vatanparvar, A.M. Shkel, J. Vac. Sci. & Technol. B 39, (2021)
32. H.T. Chua, A. Tay, Y. Wang, X. Wu, Appl. Therm. Eng. **29**, 985–997 (2009)
33. S. Tuomikoski, S. Franssila, Phys. Scr. **2004**, 223 (2004)
34. F. Niklaus, P. Enoksson, E. Kalvesten, G. Stemme, Void-free full wafer adhesive bonding. In: Proceedings IEEE Thirteenth Annual International Conference on Micro Electro Mechanical Systems (Cat. No.00CH36308), pp. 247–252 (2000)

Chapter 4
Advanced Photoresist Materials for Microsystems

This chapter explores advanced photoresist materials, including chemically amplified resists for high-resolution applications and photoresists for UV, DUV, and EUV lithography. It discusses specialty photoresists such as SU-8, HSQ, and OSTE polymers, and examines the benefits and limitations of hybrid organic–inorganic resists. Strategies for customizing photoresists for specific applications are also covered.

4.1 Chemically Amplified Resists for High-Resolution Applications

Chemically amplified resists (CARs) are a class of photoresists that are used in advanced lithography processes, including deep ultraviolet (DUV), extreme ultraviolet (EUV) and electron beam lithography. These resists use a catalytic acid-driven mechanism to obtain higher sensitivity and resolution [1]. This high-resolution patterning ability of the chemically amplified resists are the reason for their wide use in semiconductor device fabrication, MEMS and nanotechnology applications.

4.1.1 Mechanisms for Achieving Sub-Micron Features

CARs use acid diffusion control to produce sub-micron features to guarantee post-exposure delay (PED) stability and avoid feature broadening. To preserve pattern integrity and stop acid from spreading into non-exposed regions, CARs regulate acid diffusion by adding basic monomers like BCVC (3-(t-Butoxycarbonyl)-1-vinylcaprolactam) to the base polymer. Shorter acid diffusion lengths improve resolution, guaranteeing accurate feature definition even after prolonged PED durations. Copolymers of TBMA (t-Butyl methacrylate) and BCVC further improve stability

K. Wang, *Photoresist Technology in Microsystems: Principles, Processes and Applications*, Springer Series in Advanced Microelectronics 15,
https://doi.org/10.1007/978-981-95-1606-3_4

by lowering T-top formation, where the top of the resist profile becomes wider than the bottom, brought on by airborne pollutants. Scanning electron microscopy (SEM) results show that acid diffusion control is a vital mechanism in CARs for obtaining sub-micron precision, as patterns do not degrade after a 2-h delay [2] (Fig. 4.1).

CARs make use of advanced resist chemistry to obtain higher resolution at sub-micron and nano levels. Due to the shorter wavelength of the EUV light, these resists are used in EUV lithography to obtain higher-resolution patterning with decreased diffraction limitations.

By including highly absorbing elements in the resist chemistry such as Hf, Zr, Zn, and Sn, CARs improve photoacid generation efficiency, resist performance, etch resistance and feature fidelity, ensuring precise pattern transfer at the nano meter scale. This also significantly increases photoelectron production and contrast sensitivity. A study by T. Manouras and P. Argitis demonstrates how advanced CAR formulations optimize acid-catalysed hydrophilic changes and polymeric backbone cleavage, further refining contrast sensitivity, controlling acid diffusion, and enhancing pattern uniformity. Their results confirm that using EUV lithography with optimized CARs is a potent strategy for next-generation semiconductor and microsystem technologies since it allows the creation of sub-micron structures with excellent resolution and process efficiency [3]. Another study states that by increasing EUV light absorption, metal atoms in CAR formulations also maximize the usage of secondary electrons and improve photon capture efficiency. Compared to conventional organic resists, metals absorb EUV radiation more efficiently, improving resolution and sensitivity [4].

CARs also use molecular glass (MG) to improve pattern fidelity and decrease line-edge roughness (LER) in sub-micron lithography. Molecular Glass resists are made up of monodisperse, small molecular units that reduce structural fluctuations and enhance resolution, in contrast to conventional polymer resists, which have large molecular sizes and polydispersity. De Silva et al. also highlights how these resists offer a more uniform and consistent molecular spacing throughout the material, which in turn controls and regulates the acid diffusion and prevents excessive spreading, leading to sharper feature edges. Additionally, molecular glass molecules also exhibit a smaller radius of gyration (distribution of a molecule's mass around its centre of gravity) resulting in a more uniform and compact structure compared to traditional polymers, further reducing the LER. The efficacy of MG-based CARs in precisely creating sub-micron features is validated by De Silva et al.'s experimental

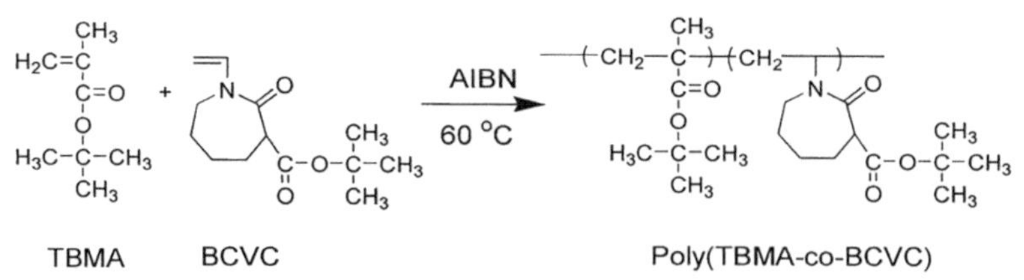

Fig. 4.1 Copolymerization of TBMA with BCVC [2]

results, which show that they consistently exhibit lower LER values than polymer resists. Utilizing these molecular characteristics, CARs with MGs offer better process stability, decreased stochastic effects, and increased resolution, which makes them perfect for applications involving next-generation lithography [5].

4.1.2 Advances in Acid-Catalysed Resist Systems

A new development in acid-catalyzed resist systems is the optimization of negative tone CARs, which allows for higher resolution in e-beam lithography. Although positive-tone resists have historically been more commonly utilized, new studies have demonstrated that negative-tone formulations can be optimized to further push resolution limitations. In order to obtain sub-40 nm features, the study develops three new experimental negative-tone CAR formulations based on Sumitomo's NEB-33 platform. Researchers outperformed prior formulations by achieving distinct 25 nm line resolutions using optimized post-exposure bake (PEB) conditions between 60–100 °C and refined photoacid generator (PAG concentrations using FTIR analysis. Furthermore, greater resolution was demonstrated by lower molecular weight resists, including HN-432, underscoring the significance of molecular weight control. These results validate the importance of negative-tone CAR optimization in obtaining sub-30 nm, higher-resolution lithographic patterning for next-generation applications [6].

In CARs, recent developments in photoacid generator (PAG) and catalyst development have greatly improved resist sensitivity, line-edge roughness (LER) control, and process stability. The research by Wang et al. presents several new anionic PAGs, such as perfluoroalkanesulfonic PAG (MTFB PAG) and methacrylate sulfonic benzene sulfonic acid PAGs (NO_2 PAG, CF_3 PAG, F_4 PAG), which enhance acid generation efficiency, DUV absorption, and resist uniformity. With higher PAG loading, faster photospeed, and less outgassing these new PAG-bound polymer resists result in lower LER and increased pattern fidelity. MTFB PAG showed the highest acidification efficiency (81%) among them, leading to better resolution and crisper feature definition [7]. These developments make PAGs perfect for high-resolution EUV lithography applications by incorporating them into the polymer backbone, which allows for more consistent acid distribution, less phase separation, and improved process control.

Single-component CARs address the inherent trade-offs between resolution, LER, and sensitivity (RLS) in EUV lithography. Single-component resists covalently bind PAGs or base quenchers to the polymer backbone, unlike conventional CARs that depend on independent PAGs and polymer matrices. This ensures more uniform acid dispersion upon exposure reducing acid diffusion variability that affect pattern fidelity, resulting in better resolution, decreased LER, and greater sensitivity.

According to the research conducted by Cen et al., these new resists are ideal for sub-10 nm patterning because they offer higher resolution and reduced swelling effects in comparison to traditional blend-type CARs. Further optimization is still necessary to overcome issues likesolubility restrictions and restricted material flexibility [8].

Another recent advancement in acid-catalyzed resist systems is the creation of three-component resists with a dissolution inhibitor, which improves sensitivity and resolution in EUV lithography. In their research, Horibe et al. provide a resist formulation that includes bis(tert-butyl) isophthalate (B-IP) as a dissolving inhibitor, triphenylsulfonium triflate as the PAG, and poly(4-hydroxystyrene) copolymerized with poly(tert-butyl acrylate) as the base resin. This system achieves high sensitivity with exposure doses below 10 mJ/cm^2, as the dissolution inhibitor initially suppresses solubility in unexposed areas but converts into a solubility promoter upon exposure, thereby enhancing feature contrast. B-IP also serves as a plasticizer, which promotes proton transport and raises sensitivity even more. The optimization of the post-exposure bake (PEB) at 110 °C guarantees higher resolution, and contrast is improved by the dissolving inhibitor, which stops the tert-butyl groups from degrading too soon. These results demonstrate that three-component CARs are a viable method for obtaining highly accurate sub-30 nm patterning [9].

Compared to conventional organic CARs, metal-oxide-based photoresists offer better etching resistance, enhanced sensitivity, and stronger EUV absorption, making them a significant development in acid-catalyzed resist systems. To improve lithographic performance, these resists use oxides of zinc (Zn), tin (Sn), hafnium (Hf), and zirconium (Zr) in different forms, such as nanoparticles, clusters, and thin films. Metal oxide resists absorb EUV photons more effectively than traditional CARs, which rely on photoacid production and solubility changes. This results in secondary electrons that start cross-linking or circuit breakdown, improving pattern resolution.

According to a study by Hasan et al., tin-oxide resists are twice as effective at absorbing EUV light as polymer-based resists, and they can reach sub-15 nm resolution at low exposure intensities (~20 mJ/cm^2). Additionally, zirconium and hafnium oxides provide up to 25 times greater etching resistance, maintaining the integrity of the design [10].

A new area of study in CARs is the creation of PFAS-free PAGs, motivated by environmental concerns about the toxicity and persistence of per- and polyfluoroalkyl substances (PFAS). PAGs including PFAS have long been a standard in semiconductor photolithography because of their excellent solubility, thermal stability, powerful acid production, and control over acid diffusion. Nonetheless, research initiatives to create non-PFAS substitutes have been prompted by sustainability objectives and regulatory pressure.

Recent research investigates PAGs that are entirely devoid of PFAS, whereas early solutions concentrated on substituting short-chain PFAS with long-chain PFAS. It is still very difficult to maintain high acid strength, thermal stability, and resist performance without PFAS. The transition to PFAS-free PAGs is expected to take 15–20 years, requiring extensive R&D, supplier collaboration, and process qualification [11].

4.2 Photoresists for UV, DUV, and EUV Lithography

4.2.1 Materials Suited for Different Lithography Wavelengths

Photolithography is a crucial technology in the semiconductor field that has developed continuously for the past 5 decades along with the decrease in wavelength to meet the demand of miniaturization of semiconductor devices [12], illustrated in Image 3.1 [13]. For each wavelength, it is required a suitable material to optimize the absorption and transmission of light in the photoresist [14]. The apperence of different used wavelenths illustrates through Fig. 4.2, based on [13].

In the early stages, UV lithography, known as g-line (436 nm-wavelength) and i-line (365 nm- wavelength) lithography, worked effectively with novolak polymer as the base resin and Diazonaphthoquinone sulfonate (DNQ) or 1,2-naphthoquinonediazide sulfonate (NQD) in specific as the photoactive compound [12, 14]. When exposed to the g-line or i-line light, the NQD compound goes through photochemical decomposition process and produces hydrophilic indene carboxylic acid. This acid would dissolve in the alkaline developer and enhance the solubility of the Novolak polymer. Consequently, the solubility between the exposed and unexposed areas would be significant and create clear patterns [14]. The resolution achieved was approximately 0.3 μm.

Along with the development of technology, the next generation of wavelengths that have been used is in deep UV, which are 248 nm (KrF excimer laser) and 193 nm (ArF excimer laser). Regarding the KrF excimer laser, the used photoresist is Polyhydroxy styrene (PHS). When the wavelength decreases to 193 nm (ArF), the Novolac polymer and PHS are not suitable anymore due to the strong light absorption of their benzene rings at around 200 nm wavelength region [14]. As a result, the base

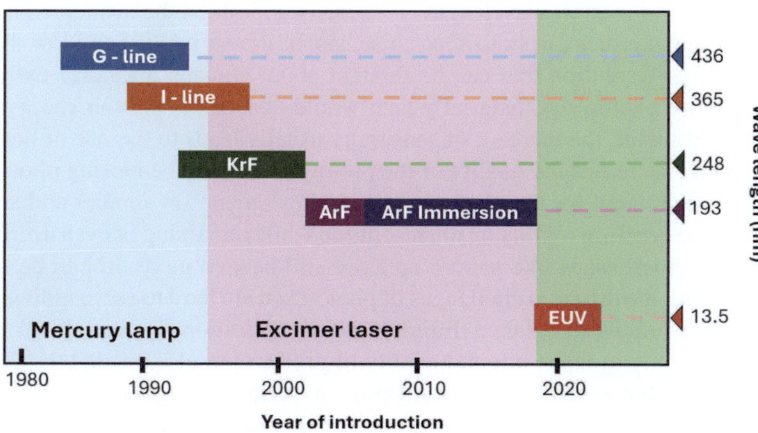

Fig. 4.2 The development of UV lithography [13]

resin chosen to replace are alicyclic polymers (saturated hydrocarbon-ring polymers), such as Adamantane and Norbornene.

Not only in deep UV but CARs are still being used in EUV (13, 5 nm) lithography nowadays. However, this kind of material is showing some limitations as the EUV lithography requires the photoresist to have higher EUV absorption, meanwhile CARs EUV light absorbing ability is low [15]. Moreover, they were limited by the high LER, which makes pattern easy to fall.

Therefore, scientists are researching other new materials that can resolve those weaknesses, and the non-CARs have attracted their attention due to their superior sample fidelity and lower LER compared to CARs [16, 17]. A typical example of this group is PMMA as it demonstrated superior resolution and LER performance [17, 18]. Besides, another proposal about the use of organometallic and coordinative materials photoresist to enhance the EUV absorption, increase sensitivity and reduce LER edge roughness was given by Gayoung [19]. To be more specific, the two types of materials mentioned are Metal Nanoclusters and Low-molecular-weight Organometallic Compounds. The Nanocluster can improve the EUV sensitivity by generating radicals through the metal–carbon bond breaking reaction. Among the experiments with photoresist containing Sn, Zn, Zr, and Hf, it was found out that Sn is the best candidate. Additionally, the Low-molecular-weight Organometallic Compounds bring the ability to control more precisely during lithography process as they contain the metal–carbon bonds that are easily decomposed upon exposure to EUV.

4.2.2 Challenges and Innovations in Extreme UV Regimes

In general, nowadays, the photoresists of EUV lithography are facing a bunch of challenges [9] due to the short wavelength (13.5 nm) that limited the available photons, which lead to the issues of photo shot noise, stochastic variability and low absorption capability. A solution that can be thought about is using a kind of extremely high sensitivity photoresist material which would absorb the photon energy more effectively. However, the increase of sensitivity usually leads to the rise of line edge roughness and decreases the quality of the prototype. Besides, choosing photoresist also has a requirement for the resolution. With the demand for smaller and smaller prototypes, the question for raising the resolution while remaining or even decreasing LER and the other issues like pattern collapse still needed more time to be solved well. Therefore, the different challenges of photoresist are hard to solve individually, they require solutions to balance different factors. Additionally, a problem that is not related directly to the result and quality of product but also needed to take into account in the photoresist improvement process is about outgassing phenomenon releases the gas from additives and decomposition products leading to the environmental pollution as well as affecting badly and decreasing the lifetime of optical system components. Limiting this phenomenon would improve the sustainability, environment and durability of the system.

Facing those challenges, the researchers are seeking for new materials that can solve those problems. As aforementioned, they are paying attention to the promise materials like non-CARs or organometallic and coordinative to replace the traditional CARs photoresist. Their main focus now is enhancing photo absorption, designing ordered resist systems and improving solvents as well as developers to enhance pattern accuracy.

4.3 Specialty Photoresists for Microsystems

4.3.1 SU-8: High-Aspect-Ratio Capabilities and Common Uses

SU-8 is a specially designed negative photoresist that contains epoxy groups. When exposed to UV light, these epoxy groups undergo cross-linking, creating new bonds between polymer chains. The cross-linked regions become insoluble in the developer solution, making them more resistant to etching. This property is particularly useful in applications requiring high etch resistance, such as fabricating high aspect ratio structures [20, 21].

High aspect ratio structures have to undergo a long etching process to achieve the desired depths, so negative photoresists with crosslinked agents is the preferable selection for photoresist for this application. Negative photoresist can be applied in thicker layers without compromising pattern fidelity, while crosslinking agents enhance the mechanical and chemical strength of the resist. These thicker and durable layers of photoresist are more robust during the etching process helping create deep, steep-sided features [22].

Some of the applications that require a high aspect ratio, and consequently the use of SU-8, include MEMS, microfluidics, and super-hydrophobic surfaces.

MEMS devices, such as accelerometers, gyroscopes, and comb drive actuators, often incorporate high aspect ratio (HAR) structures. Similarly, microfluidics utilize HAR features like deep channels, tall pillars, or complex 3D geometries, with SU-8 being a preferred material due to its biocompatibility. Additionally, super-hydrophobic surfaces depend on HAR structures, such as pillars or needles, to achieve their water-repellent properties [22].

Three key steps that need to be carefully optimized are exposure, post-exposure bake (PEB), and development to achieve the best results, considering the specific properties of SU-8.

Exposure-UV or EUV exposure triggers acid production, which diffuses more at higher radiation energy levels, especially with prolonged exposure. To prevent underexposure and incomplete cross-linking-particularly at the bottom of the resist-the dose must be carefully controlled. An improper dose increases acid diffusion, reducing pattern fidelity. Therefore, the dose should be optimized based on resist thickness to ensure proper feature definition while minimizing acid diffusion before the PEB step.

For thinner layers (<10 μm), lower doses (100–150 mJ/cm^2) are used to minimize diffusion. For medium layers (10–50 μm), moderate doses (up to 300 mJ/cm^2) ensure proper cross-linking and resolution. For thicker layers (>50 μm), higher doses (up to 800 mJ/cm^2) are needed to maintain pattern fidelity [23].

Post-exposure bake (PEB)—This step ensures cross-linking in the photoresist after exposure by activating the acid and catalyzing the ring-opening reaction. It also traps the acid, preventing further diffusion. Higher temperatures harden the structure but also introduce internal stress. Lower temperatures, on the other hand, reduce stress but may result in incomplete cross-linking, lowering etch resistance. 90 °C is usually chosen as the ideal temperature for balancing resolution and mechanical strength.

Development—To achieve the desired results, the right developer, development time, and temperature must be carefully selected. This helps prevent roughness, lift-off, and swelling. PGMEA or cyclopentanone are the most commonly used developers for SU-8. The process is typically performed at room temperature (20–25 °C) for 1–5 min [24].

4.3.2 *Hydrogen Silsesquioxane (HSQ): Resolution and Etch Resistance*

Hydrogen Silsesquioxane (HSQ) is a type of negative photoresist well known for high-resolution patterning capabilities and etch resistance. It consists of a network of hydrogenated silicon-oxygen (SiO_1-$_5H$) units, which provide a rigid structure. Upon exposure to an electron beam, the Si–H bonds break and form new Si–O-Si bonds through a crosslinking reaction. This transforms the material into a structure resembling SiO_2, which is insoluble in common alkaline developers. This property grants the resist excellent etching resistance, making it particularly useful for low-resolution applications where robust patterning is required.

Its inorganic nature ensures that when exposed to high energy, it neither expands nor deforms, preserving the pattern integrity. This is particularly critical for smaller-scale applications, where even minor distortions can significantly impact performance.

All the elements in the composition (Si, O, and H) have low atomic numbers which results in less electron backscattering. The reduced electron backscattering reduces unintended exposure, improving the resolution [25].

Here is a comparative analysis between SU-8 and HSQ, highlighting their advantages and disadvantages in terms of resolution, aspect ratio, etch resistance, and cost (Table 4.1).

Therefore, HSQ is used when high resolution is required for nanoscale patterning and when high etch resistance is needed, such as in deep etching with RIE.

Table 4.1 Comparative analysis between SU-8 and HSQ

Criterion	SU-8	HSQ	Best choice
Resolution	High (~0.5 μm)	Very high (~10 nm)	HSQ
Aspect ratio	Excellent (>20:1)	Low	SU-8
Etch resistance	Moderate to high	Very high	HSQ
Cost	Lower	Higher	SU-8

4.3.3 OSTE Polymers: Customization and Biocompatibility

Specialty photoresist materials called OSTE (Off-Stoichiometry Thiol-Ene) polymers are frequently used in microsystem applications, especially in lab-on-a-chip (LoC), BioMEMS, and microfluidic devices. The high-resolution patterning capacity of OSTE polymers, in contrast to traditional photoresists, makes them ideal for more precise microfabrication. Their capacity to be microstructured via photopatterning, direct lithography, and soft lithography enables the highly precise creation of complex micro-scale structures [26].

Furthermore, OSTE's thiol-excess characteristic guarantees structural integrity even for multi-layered microfluidic devices by enabling unassisted low-temperature bonding [26]. They have two-step curing procedure, which enables exact control over surface chemistry, mechanical properties, and bonding characteristics making them customizable. The stiffness and elasticity of OSTE polymers can also be controlled by varying the stoichiometric ratios, which enables them to be used for both soft, flexible features and hard structural components [27].

OSTE polymers also make surface modification and sensor integration simple, making it easier to attach coatings, biomolecules, or electronic components due to their reactive surface chemistry. Advanced $OSTE^+$ polymers improve biocompatibility and have cell compatibility comparable to that of polystyrene and polycarbonate, which makes them perfect for BioMEMS and biomedical applications. $OSTE^+$ also provides resistance to hostile chemical environments and high-temperature stability (Tg > 115 °C), guaranteeing long-term durability under demanding working circumstances. Additionally, OSTE polymers allow for adhesive-free attachment to a variety of substrates, such as silicon, glass, and plastics, facilitating the smooth integration of both rigid and flexible parts in next-generation biomedical and microfluidic systems [27].

4.3.4 Dry Film Photoresists: Applications in Roll-To-Roll Processing

Dry film photoresist is the type of photoresist that is applied to substrates in a solid, pre-coated film form and originally used for PCB. It has many advantages as well as characteristics that make it suitable for roll-to-roll application. Firstly, the DFR process only requires a simple lamination step without a high temperature requirement as the performing temperature is below 80 °C. So that it can also be used for most polymeric substrates. Moreover, this method does not require any post-annealing process anymore. Therefore, it is not only easy, fast and convenient when applied in the roll-to-roll manufacturing process but also decreases the cost and increases productivity. Those points had been proved and illustrated by Hwang and his team through the Ag nanowire flexible electrodes manufacturing process [28]. He also mentioned that, in the roll-to-roll process, especially for effective Ag nanowires manufacturing process, patterning is a vital step. If the method applied is liquid photoresist, there would be some challenges like non-uniform thickness of photoresist that require a post-annealing thermal step. This would be a limitation of liquid photoresist compared to dry photoresist in roll-to-roll fabrication processes. Besides, another notable point is that DFR process is friendly to the environment due to the alkaline development solvents it uses are less toxic compared to the organic solvents of liquid photoresist process, which would improve the sustainability-an important factor when manufacturing electronic products. Overall, in the applications for roll-to-roll process, dry film photoresist is a suitable method that is fast, cheap, easy and convenient in many aspects.

4.3.5 Hybrid Organic–Inorganic Resists: Benefits and Limitations

Wang conducted research about Organic–inorganic Hybrid Photoresists [29] and give a positive assessment of these materials. These hybrid photoresists combined both strong points of organic and inorganic materials, which make them become promising for EUV lithography. Firstly, the Organic–inorganic Hybrid Photoresists have the high EUV absorption and etching resistance of inorganic materials. Besides, the organic components contained in hybrid photoresists can be changed to enhance the characteristics such as sensitivity and solubility. Moreover, they also have good chemical and thermal stability. However, these hybrid photoresist materials are still facing limitations and challenges. Similar to inorganic materials, the Organic–inorganic Hybrid Photresists have low sensitivity, making them unsuitable for commercial applications. Some of the materials discussed in the paper also met their specific issues, such as the increase in cost of nanoparticle/metal oxo-cluster and organometallic complex photoresists. Lastly, there are many remaining questions including the structures of the insoluble substant, the function of secondary electrons and chemical characteristics, that required further research.

4.4 Customizable Photoresists for Specific Applications

A single photoresist formulation cannot meet the requirements of all applications. Depending on what is being fabricated and its specific requirements, different application-specific properties are required for the photoresist. Some of the factors that determine these specific requirements for photoresist are resolution, aspect ratio, lithography method, process method, and biocompatibility [30].

4.4.1 Strategies for Tailoring Photoresist Properties to Application Needs

As the resolution increases (smallest feature size decreases) we need to maintain sharp, well-defined features. Higher resolution introduces issues such as resist edge variations, increased light absorption at smaller wavelengths, and the need for ultra-thin resists that retain structural integrity.

To achieve high resolution in a lithographic process, a photoresist must have high contrast, low absorption, thin films, and carefully controlled sensitivity.

High contrast—To keep high well-defined patterns, sharp features and less roughness in the edges, we need high contrast resist. Contrast is a resist property that defines how it differentiates between exposed vs unexposed areas. High contrast means photoresist reacts very sharply to light exposure, creating clear, well-defined and sharp patterns.

Several methods can enhance the contrast of a photoresist. In positive resists, increasing the photoactive compound (PAC) concentration improves solubility, leading to better-defined patterns. Modifying the polymer matrix (mechanical structure, chemical properties, and light response) can help slow down the movement of acid generated during exposure, preventing excessive diffusion that blurs feature edges [31].

Low absorption—In general photoresist should have controlled absorption properties to allow radiation energy to penetrate deeper to get exposure in lower layers. If bottom layers are underexposed there will be fluctuations in feature edges. At smaller scales, these variations become more critical. High absorption also causes more waves to bounce between substrate and resist, which causes unintended exposure and pattern distortion.

To reduce the absorption of the photoresist in EUV and DUV light, fluorinated polymers should be used [31].

Thin film resist—Thick layer of resist does not allow light to penetrate uniformly. In EUV and DUV lithography, resist absorbs more light. Light attenuates as it travels deeper meaning the top layer gets more exposure than the bottom. This effect, similar to high absorption explained above, causes fluctuations in feature edges. The thicker the layer of photoresist, the greater the exposure variation.

To ensure uniform light penetrating thin films of photoresist should be applied [32].

Controlled sensitivity—Sensitivity explains how efficiently resist responds to exposure. If a resist has high sensitivity, it means that small doses of exposure generate a large chemical reaction. Small amount of light generates a large acid concentration which then diffuses beyond the exposed region. This causes edge blurring and a reduction in resolution.

Low sensitivity on the other hand slows down the process, since it requires more exposure and time to create the pattern [32]. In the industry they aim for a sweet spot where sensitivity is neither too low (causing slow processing) nor too high (leading to feature blurring). They achieve this balance by controlling acid diffusion, optimizing contrast, and choosing medium sensitivity [33]. For high aspect ratio structures, we need to consider photoresist properties like etch resistance, mechanical strength, and adhesion to substrate. Etch resistance—In high aspect ratio structures, to achieve the desired depths, it has to undergo a long etching process. The photoresist has to withstand this long etching process without degrading and serve as a durable mask.

To enhance etch resistance, two methods are mainly used—adding compounds with multiple carbon rings to increase carbon content and ring density, or an additional step that takes place after pattern development where you expose the resist to gas vapors that results in deposition of aluminum oxide within the resist structure [34].

Adhesion—The resist layer must remain intact on top of the structure throughout the process to prevent deformities during etching, as it defines the shape. Therefore, it is crucial that this layer does not peel off the substrate. This can be controlled by selecting photoresists with strong adhesion to the substrate.

Adding adhesion—Promoting agents into the formulation of the photoresist will improve the adhesion to various substrates. The chemical structure of the photoresist should be compatible with the substrate. Different substrates (e.g., silicon, glass, metals) have varying surface chemistries that can affect photoresist adhesion. Choosing a resin that interacts favorably with the substrate material can enhance adhesion [31].

Mechanical strength—during etching of high aspect ratio structures several mechanical stresses can impact structure integrity. Some of them are capillary forces caused during drying that pull features can cause collapse of the structure, adhesion forces can cause stiction, and also thermal expansion difference between substrate and photoresist can cause stresses and eventually cracks [31].

Therefore, photoresist mechanical properties such as mechanical strength (photoresist with a higher Young's modulus exhibits greater stiffness, photoresists that form densely cross-linked polymer networks upon exposure create a rigid structure, photoresists that are engineered to withstand elevated temperatures), and thermal expansion compatibility should be selected to prevent bending and withstand stresses [33] (Fig. 4.3).

Different lithography technologies use different sources to pattern the photoresist. Selection of photoresist should be according to the exposure source used. Optical lithography—It uses UV light exposure to pattern the substrate. Therefore, photoresist used in this case should be sensitive to specific wavelengths of UV light applied. In

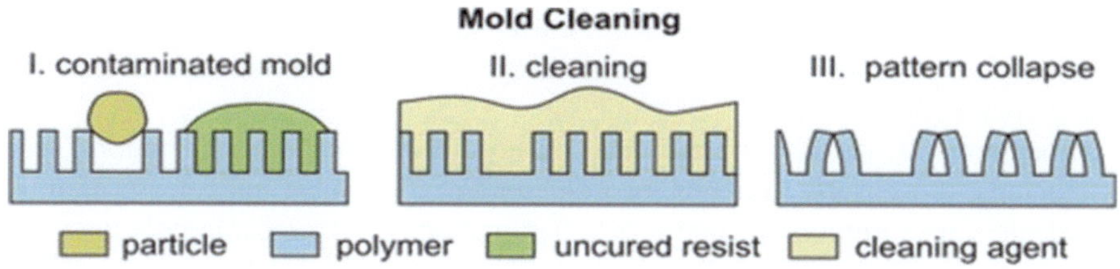

Fig. 4.3 Capillary forces during the cleaning of polymeric molds can lead to structural collapse [35]

today's industry, the primary ultraviolet (UV) wavelengths utilized are 193–13,5 nm. Resists must be tailored for UV sensitivity, chemical amplification, and high resolution. For UV sensitivity at larger wavelengths, alicyclic compounds should be added to resist to ensure the right transparency, while for smaller wavelengths heavy elements should be added to increase absorbance [33].

Electron Beam Lithography (EBL)—The chemical reaction in the photoresist is driven by electrons that hit the surface and generate secondary electrons. The photoresists used in this technology should be polymers that undergo chain scission or crosslinking upon electron exposure (electrons causes the polymer chains to break into smaller fragments), like polymethyl methacrylate (PMMA) and hydrogen silsesquioxane (HSQ).

By Incorporate chemical groups or additives like halogenated compounds or heavy elements to traditional resist, we can enhance electron sensitivity [36]. Nanoimprint Lithography (NIL)-This technology relies on mechanical molding on the pattern, instead of exposure. So, the stamp used is pressed into the resist that is applied on the substrate, then the resist is cooled or exposed to UV to harden. The resist should have low viscosity and be able to deform easily [37]. Beyond lithography, other processes like etching, deposition, doping, and bonding also affect the selection of resist properties.

Etching-For wet etching, selecting a photoresist with the right chemical resistance to acids or bases, and also strong adhesion to the substrate to prevent undercutting is important. For dry etching, the resist must have high resistance to plasma gases and the ability to withstand high temperatures.

Deposition-For physical deposition, resist should have the ability to withstand high temperatures, and adhesion to substrate.

For chemical deposition, resist should have the ability to chemically resist the reactive gases. This can be achieved by using more stable polymers like fluoropolymers, polyimides, or silicon-containing polymers, or by incorporating inorganic additives [34].

Doping-Depending on which doping method is used, ion implantation or diffusion, different resits properties are needed. For ion implantation, polymers with higher molecular weight are needed because they are harder to break, preventing the masked areas by high ion energy. While for diffusion, the resist should be able to withstand higher temperatures without melting, preventing diffusion of impurities in the masked areas. This can be achieved by selecting polymers with high glass transition temperatures or incorporating additives like silicon dioxide or ceramic nanoparticles [31].

Biocompatibility - Photoresists that are used in biomedical applications such as biochips, biosensors, implants, or tissue engineering must not release harmful chemicals, cause immune response or inflammation. Biodegradable polymers like polylactic acid (PLA) or polyglycolic acid (PGA) should be used [38].

Optimizing photoresist properties such as sensitivity, resolution, mechanical strength, LER, and thickness often involves trade-offs, as improving one property can negatively affect another. For example, increasing sensitivity can reduce resolution due to acid diffusion, while improving mechanical strength can increase line-edge roughness. Therefore, these properties must be carefully balanced to meet the specific demands of the application.

Sensitivity vs. Resolution—For mass production, where faster processing is needed, sensitivity is prioritized. For nanostructures, where pattern fidelity is critical, resolution is given more importance. For MEMS and microfluidics, where moderate resolution and high structural strength are required, resolution and sensitivity are carefully balanced [39].

Resolution vs Etch resistance – high etch resistance is related to high crosslinking, which reduces the ability for nanostructures. For EUV lithography, higher resolution is prioritized, which requires lower cross-linking and results in lower etch resistance. For RIE (Reactive Ion Etching), etch resistance is prioritized over resolution to protect patterns during deep etching [39].

Sensitivity vs. LER (Line-Edge Roughness)—For logic chips LER is prioritized over sensitivity. This is achieved by controlling acid diffusion. For memory chips, sensitivity is prioritized to reduce production costs, while for high-aspect-ratio structures, sensitivity is prioritized to ensure complete pattern formation and structural integrity [39].

4.4.2 Role of Dopants and Crosslinking Agents

Plasma etching causes the photoresist to degrade, which results in deformities in the pattern that it's supposed to transfer in the substrate. This is more important in the transition areas (openings of the mask). On the other hand, when photoresist is exposed to ion radiation, the high energy causes formation of a carbon-rich layer which is more resistant to etching. The ion bombardment breaks chemical bonds, forming new bonds between polymer chains that are resistant to solvents. This etch resistance is exactly what's needed in the mask to support the transfer of the pattern accurately in the substrate [21] (Fig. 4.4).

A similar result can be achieved by modifying the properties of the photoresist. Crosslinking agents incorporated in photoresist, under UV exposure start creating covalent bonds between polymer chains. This alters the structure of the photoresist

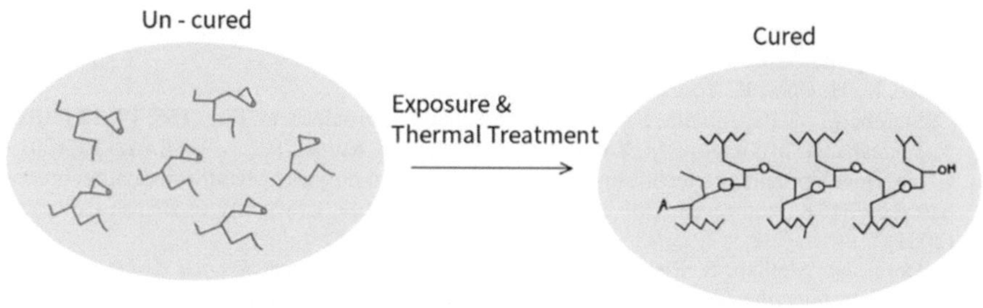

Fig. 4.4 Crosslinking effect in photoresist after radiation

differently from what exposure typically would, by changing its solubility (making it insoluble in the developer solution) [21].

This phenomenon is specifically used for high aspect ratio structures, where etch resistance and mechanical strength is needed in the photoresist. For this application crosslinking agents in negative photoresist are used. While effectively facilitating the transfer of fine structural patterns, this approach also introduces the challenge of removing the photoresist, which now has altered solubility, potentially leaving residue behind.

References

1. C.A. Mack, 614, 56–68 (1995)
2. J.-B. Kim, J.-H. Choi, Y.-G. Kwon, M.-H. Jung, K.-H. Chang, Polymer **40**, 1087–1089 (1999)
3. T. Manouras, P. Argitis, *Nanomaterials (Basel)* 10, (2020)
4. D. De Simone, I. Pollentier, G. Vandenberghe, J. Photopolym. Sci. Technol. **28**, 507–514 (2015)
5. A. De Silva, N.M. Felix, C.K. Ober, Adv. Mater. **20**, 3355–3361 (2008)
6. J. Saint-Pol, S. Landis, C. Gourgon, S. Tedesco, R. Hanawa, M. Suetsugu, M. Akita, S. Yamamoto, Microelectron. Eng. **67–68**, 274–282 (2003)
7. M. Wang, W. Yueh, K.E. Gonsalves, Macromolecules **40**, 8220–8224 (2007)
8. J. Cen, Z. Deng, S. Liu, Polym. Chem. **15**, 4599–4614 (2024)
9. H. Horibe, K. Ishiguro, T. Nishiyama, A. Kono, K. Enomoto, H. Yamamoto, M. Endo, S. Tagawa, Polym. J. **46**, 234–238 (2014)
10. M.W. Hasan, L. Deeb, S. Kumaniaev, C. Wei, K. Wang, *Micromachines (Basel)* 15, (2024)
11. S.P.C.P.W. Group, (2023)
12. T. Ito, S. Okazaki, Nature **406**, 1027–1031 (2000)
13. X. Wang, P. Tao, Q. Wang, R. Zhao, T. Liu, Y. Hu, Z. Hu, Y. Wang, J. Wang, Y. Tang, H. Xu, X. He, Mater. Today **67**, 299–319 (2023)
14. A. Soyano, Nippon Gomu Kyokaishi **85**, 33–39 (2012)
15. B. Wu, A. Kumar, Journal of vacuum science & technology b: microelectronics and nanometer structures processing. Meas., Phenom. **25**, 1743–1761 (2007)
16. A.M. Goethals, R. Gronheid, F. Van Roey, H.H. Solak, Y. Ekinci, J. Photopolym. Sci. Technol. **19**, 501–506 (2006)
17. R. Gronheid, H.H. Solak, Y. Ekinci, A. Jouve, F. Van Roey, Microelectron. Eng. **83**, 1103–1106 (2006)
18. H.H. Solak, Y. Ekinci, P. Käser, S. Park, Journal of vacuum science & technology b: microelectronics and nanometer structures processing. Meas., Phenom. **25**, 91–95 (2007)

19. G. Lim, K. Lee, S. Choi, H.J. Yoon, *Coordination Chemistry Reviews* 493, (2023)
20. A. Lio, Synchrotron Radiation News **32**, 9–14 (2019)
21. J. Lee, K.-H. Choi, K. Yoo, Micromachines **6**, 1–18 (2014)
22. C. Steinberg, M. Papenheim, S. Wang, H.-C. Scheer, Microelectron. Eng. **155**, 14–18 (2016)
23. V. Vlnieska, A. Mikhaylov, M. Zakharova, E. Blasco, D. Kunka, *Polymers (Basel)* 11, (2019)
24. C. Ge, Development of a technology framework for rapid polymer MEMS fabrication process based on SU-8 photoresist, University of British Columbia, University of British Columbia, (2021)
25. V. Desai, M. Mellish, S. Bennett, N.C. Cady, *Journal of Vacuum Science & Technology B, Nanotechnology and Microelectronics: Materials, Processing, Measurement, and Phenomena* 35, (2017)
26. T. Haraldsson, C.F. Carlborg, W. van der Wijngaart, *Proc Spie* 8976, (2014)
27. C. Errando-Herranz, A. Vastesson, M. Zelenina, G. Pardon, G. Bergström, W.v.d. Wijngaart, T. Haraldson, H. Brismar, K.B. Gylfason, Biocompatibility of OSTE polymers studied by cell growth experiments, In: 17th International Conference on Miniaturized Systems for Chemistry and Life Sciences, Freiburg, Germany, (2013)
28. B. Hwang, P. Matteini, *Fashion and Textiles* 9, (2022)
29. Z. Wang, X. Yao, H. An, Y. Wang, J. Chen, S. Wang, X. Guo, T. Yu, Y. Zeng, G. Yang, Y. Li, Journal of Microelectronic Manufacturing **4**, 1–8 (2021)
30. R.C. Jaeger, (1988)
31. M.J. Madou, (2018)
32. B.W. Smith, K. Suzuki (2018)
33. R.C. Ramola, S. Chandra, Defect and Diffusion Forum **341**, 69–105 (2013)
34. M.R. Krishnan, E.H. Alsharaeh, *Journal of Polymer Science and Engineering* 6, (2023)
35. A. Finn, B. Lu, R. Kirchner, X. Thrun, K. Richter, W.-J. Fischer, Microelectron. Eng. **110**, 112–118 (2013)
36. A.S. Gangnaik, Y.M. Georgiev, J.D. Holmes, Chem. Mater. **29**, 1898–1917 (2017)
37. Z. Jia, J. Choi, S. Park, Nanotechnology **29**, 365302 (2018)
38. T. Zahra, U. Javeria, H. Jamal, M.M. Baig, F. Akhtar, U. Kamran, Anal. Chim. Acta **1316**, 342880 (2024)
39. K. Karimi, A. Fardoost, N. Mhatre, J. Rajan, D. Boisvert, M. Javanmard, *Micromachines (Basel)* 15, (2024)

Chapter 5
Photoresist Processing Techniques

Photoresist processing lies at the heart of modern microfabrication, enabling precise transfer of intricate patterns onto substrates. This chapter delves into the comprehensive suite of techniques employed in photoresist processing, from initial coating to final treatments, providing a holistic view of the fabrication workflow. First, we explore the critical coating methods; Spin, spray, and dip coating, that dictate film uniformity and quality. We then move onto the exposure phase, examining both mask-based and maskless techniques, emphasizing recent advances in laser and electron-beam lithography. Subsequent sections cover development processes, contrasting wet and dry methods, with insights into the chemistry of developer solutions and their impact on feature resolution. Finally, we address post-exposure and post-development treatments (annealing and hardening), that enhance the durability and adhesion of the patterned structures.

5.1 Photoresist Coating Techniques: Spin, Spray, and Dip Coating

Photoresist coating is crucial for semiconductor fabrication, enabling intricate patterning in microelectronics, optics, and nanotechnology. Spin coating, pioneered in mid-twentieth century, revolutionized integrated circuit manufacturing with ultra-thin uniform films. Its limitations include substrate constraints and material waste spurred innovations in spray, dip, and niche methods like roll and blade coating. This review examines these techniques, their principles, parameters, and performance, addressing emerging trends and sustainability challenges.

K. Wang, *Photoresist Technology in Microsystems: Principles, Processes and Applications*, Springer Series in Advanced Microelectronics 15,
https://doi.org/10.1007/978-981-95-1606-3_5

5.1.1 Spin Coating

Principle: Centrifugal force distributes photoresist across flat substrates, driven by precise control of spin speed, viscosity, and environmental conditions.

Working Mechanism:

Substrate preparation is critical for adhesion, requiring the removal of contaminants such as dust or oil, which is done by solvents or plasma cleaning [1]. Once the substrate is clean, photoresist is dispensed onto it, followed by high-speed rotation (1,000–6,000 rpm) that uses centrifugal force to spread the resist evenly across the wafer surface. After spinning, part of the solvent evaporates from the photoresist, leaving behind a uniform thin film. The process concludes with a final curing step that further dries and stabilizes the film, which also helps increase the overall process yield [1].

Key Parameters and Optimization Strategies

The spin coating process is governed by several key parameters that significantly influence film quality. Spin speed (ω) has an inverse relationship with the resulting film thickness, defined by the equation;

$$h = 0.94 \cdot \left(\frac{\eta U}{\gamma}\right)^{\frac{2}{3}} * \left(\frac{\eta}{\rho g}\right)^{1/6}$$

where h represents thickness, η denotes viscosity, and k, β, and p are empirically determined constants [2]. Viscosity (η) exhibits a direct proportional relationship with film thickness, making higher viscosity formulations particularly suitable for thicker films such as those used in nanoparticle coatings [3]. The acceleration profile must be carefully controlled, as gradual acceleration prevents defects including streaks or bubbles that compromise film integrity [1]. Edge bead formation, the accumulation of excess material at substrate peripheries, requires mitigation through optimized spin profiles or dedicated post-coating removal techniques [4].

Several optimization strategies have emerged to enhance spin coating performance. Self-leveling resins have been developed to achieve remarkable thickness uniformity, with variations below 1% through inherent fluid dynamic properties [5]. Meanwhile, modern fabrication facilities increasingly incorporate real-time monitoring systems, to continuously assess resist thickness and uniformity throughout the process [6].

Applications:

The versatility of spin coating enables its application across diverse fields, including microelectronics (particularly photolithography), optical lens fabrication, and biomedical device production [6].

5.1.2 Spray Coating

Spray coating atomizes photoresist into micrometer-sized droplets for deposition on large or textured substrates, offering solutions where conventional techniques fall short. This approach has gained significance in industrial applications, particularly for organic thin-film transistors, solar cells, and components with irregular geometries where spin coating proves inadequate [7].

Working Mechanism

The process begins with atomization, where nitrogen-assisted nozzles or ultrasonic devices break the photoresist solution into fine droplets [7]. These droplets are then transported by controlled gas flow to the substrate surface. During this transformation, partial solvent evaporation occurs, which contributes to the technique's exceptional ability to provide coverage on textured surfaces and irregular edges [7]. This in-transit evaporation distinguishes spray coating from other methods and enables its unique capabilities for complex geometries.

Key Parameters and Optimization Strategies

Spray pressure significantly influences the final coating quality, with higher pressures generally refining droplet size and enhancing uniformity across the substrate [7]. The distance between the nozzle and substrate requires careful calibration, with optimal ranges typically falling between 10–30 cm to balance adequate coverage and controlled solvent evaporation [7].

Solvent composition plays a crucial role in film formation dynamics. Fast-evaporating solvents like acetone can reduce edge buildup issues, though they demand precise environmental humidity control to prevent coating defects [7]. Substrate temperature management, typically between 50–80 °C, improves film uniformity by accelerating solvent evaporation and promoting more consistent film formation [7].

Advanced equipment may employ mixed nozzle systems that combine internal and external atomization mechanisms to homogenize droplet distribution across the substrate. The importance of a particle-free environment cannot be overstated, with most industrial systems incorporating HEPA filters and laminar airflow arrangements to minimize contamination risks [7].

Applications:

Spray coating is particularly valuable for large-area electronics, textured solar cells, and automotive coatings where conventional methods face significant limitations [7].

5.1.3 Dip Coating

Dip coating represents one of the most straightforward yet versatile methods for photoresist application. This technique operates on the fundamental principles of substrate immersion and controlled withdrawal from a photoresist bath, effectively leveraging viscous drag and capillary action to create uniform films.

Working Mechanisms

The process begins with complete immersion of the substrate into the resist solution, allowing thorough wetting of all surfaces. The subsequent controlled withdrawal phase is where the magic happens-as the substrate emerges from the solution, a thin liquid film adheres to the surface. This withdrawal stage critically determines the final film characteristics according to the Landau-Levich equation:

$$h = 0.94 \cdot \left(\frac{\eta U}{\gamma}\right)^{\frac{2}{3}} * \left(\frac{\eta}{\rho g}\right)^{1/6}$$

where $\gamma =$ surface tension, $\rho =$ density, and $g =$ gravity.

Following withdrawal, a carefully controlled drying phase occurs in a saturated atmosphere to minimize defects through gradual solvent evaporation.

Key Parameters and Optimization Strategies

Several parameters significantly influence the quality of dip-coated films. Withdrawal speed (U) exhibits direct proportionality to film thickness, with higher speeds yielding thicker coatings. Solution viscosity (η) plays a dual role-higher viscosity improves adhesion properties but necessitates slower withdrawal rates for optimal results. Maintaining proper solvent saturation in the drying environment prevents premature evaporation that could lead to cracking or other film defects [8].

Optimization of the dip coating process often involves using low-boiling solvents to strategically increase resist viscosity, providing better thickness control throughout the manufacturing process [8]. Equally important is precise immersion rate control, which ensures uniform wetting during the initial substrate entry phase, preventing air entrapment and resultant coating irregularities [8].

Applications:

Dip coating's industrial value lies in its low cost and versatility for applications like optical and decorative films. Its ability to produce uniform anti-reflective coatings and its suitability for roll-to-roll manufacturing are key advantages.

5.1.4 Emerging and Niche Techniques

Roll Coating

Roll coating applies photoresist via precision rollers, making it particularly suitable for flexible substrates including paper, plastics, and metallic foils. This technique achieves remarkable material utilization efficiency of approximately 95%, while maintaining high uniformity across large surfaces and seamless integration with continuous manufacturing processes [9]. The versatility of roll coating has expanded its application across diverse industries including electronics, packaging, automotive, textiles, and energy sectors, with growing demand in several of these fields. Despite these advantages, roll coating remains primarily limited to macroscale uniformity applications, typically producing layers no thinner than 1 μm [9].

Inkjet Printing

Inkjet printing enables direct patterning of photoresist with resolution between 20–30 μm [10], offering a cost-effective solution, particularly valuable in research and development contexts. The technology provides precise material deposition without requiring masks, significantly reducing the number of process steps compared to conventional methods. Modern industrial inkjet systems have evolved to accommodate a wide range of functional materials, including conductive inks, biological materials, and specialty polymers [10]. This versatility creates substantial advantages in customization and rapid prototyping scenarios, with applications extending into electronics, pharmaceuticals, and tissue engineering. However, the technology's relatively low throughput continues to limit its suitability for mass production environments [10].

Chemical Vapor Deposition (CVD)

Chemical Vapor Deposition facilitates gas-phase deposition of conductive and dielectric layers with exceptional precision. The technique excels at non-line-of-sight coating for geometrically complex substrates and can effectively coat high-aspect-ratio structures that would challenge traditional liquid-based methods [11]. Recent breakthroughs in atmospheric pressure and plasma-enhanced processes have enabled CVD operations at significantly lower temperatures, expanding compatibility to temperature-sensitive materials such as polymers and textiles. Despite these advances, CVD implementation typically requires substantial infrastructure investment, including high-vacuum systems and cleanroom facilities [11].

5.1.5 Comparative Analysis

See Table 5.1.

Selection Guidelines:

Spin Coating: Flat substrates (silicon wafers), nanometer precision [1].

Spray Coating: Textured/large substrates (solar panels), rapid deposition [7].

Dip Coating: Roll-to-roll processes, cost-sensitive applications [8].

Roll/Blade Coating: High-throughput industrial coatings (paper, foils) [9].

Table 5.1 Performance metrics of coating techniques

Metric	Spin coating	Spray coating	Dip coating	Roll/Blade coating
Thickness	< 1% variation	Moderate (~5%)	Moderate (~5%)	Low (~10%)
Uniformity	< 1% variation	Moderate (~5%)	Moderate (~5%)	Low (~10%)
Material utilization	Low (<10%)	High (~90%)	Moderate (~50%)	Very High (~95%)
Adhesion strength	High (with cleaning)	Medium (needs surface treatment)	Medium	High
Throughput	Medium (lab-scale)	High (large area)	Medium (batch)	Very High (continuous)
Cost-effectiveness	High (precision)	Moderate (equipment)	Low (simple setups)	High (industrial)

Table 5.2 Common spin coating issues

Issue	Root cause	Solution
Edge bead formation	Centrifugal force	Edge bead removal (EBR) systems
Air bubbles	Trapped air during dispensing	Degas resin pre-deposition
Particle contamination	Dirty environment	Cleanroom protocols, HEPA filters
Thickness variation	Inconsistent spin speed	Closed-loop speed control

Challenges and Solutions:

See Table 5.2.

5.2 Exposure Methods: Mask-Based and Maskless Techniques

In this section, we will explore in depth the exposure of a coated wafer and its effect on the photolithography process. Exposure is the step where we expose the coated wafer to a light source to build up or transfer the pattern. This step is used to project or transfer the pattern onto the photoresist. Only a certain portion of the photoresist is exposed to light according to the design of the mask [12].

In the exposure methods, we find two major approaches for exposure. Illustrated in Fig. 5.1 are mask-based and maskless lithography. Each has its own advantages and disadvantages. Mask-based lithography relies on prefabricated masks to design patterns and define them. On the other hand, maskless approach utilizes direct writing techniques, allowing for greater flexibility and cost-effectiveness. Let's delve into the details in the following sections.

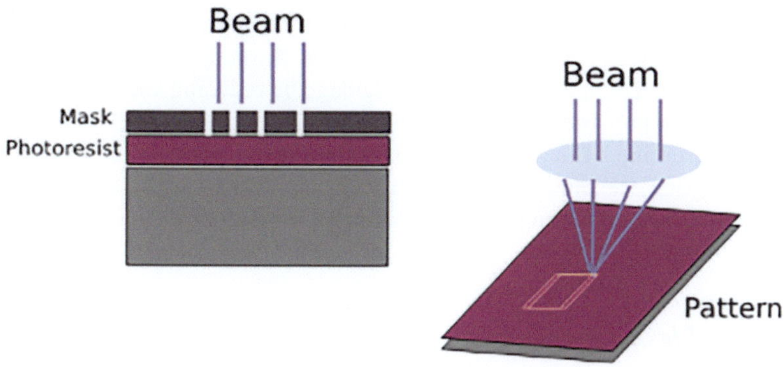

Fig. 5.1 Mask-based (left) and maskless (right) exposure techniques

5.2.1 Mask-Based Techniques

A mask-based technique involves the use of photomask to project patterns onto the photoresist layer. The photomask is a transparent plate with a patterned opaque layer that blocks some light.

The process begins with the application of a photoresist layer to the substrate, which is then exposed to light through a photomask. A mask aligner is utilized to align the coated wafer with a mask. The areas of the photoresist exposed to light undergo a chemical change, allowing the exposed or unexposed regions (depending on whether the photoresist is positive or negative) to be washed away in a subsequent development step [13].

The pattern from the photomask is thus transferred to the photoresist and, subsequently, to the underlying substrate. This method is well-suited for high-volume production as the mask can be reused many times.

When the coated substrate is exposed to light, several lithography exposure modes are available for use. The operator can carefully adjust the distance between the mask and the wafer, known as the "exposure gap," using the mask aligner. Altering this gap affects both the resolution of the pattern on the substrate and the overall efficiency of the manufacturing process. The exposure modes are divided into two categories: contact and proximity exposure. The most frequently utilized are contact exposure modes.

In contact mode, the mask and wafer are brought together and then positioned at the alignment gap before being pressed together for light exposure. This method includes various exposure modes, each affecting resolution and substrate patterning integrity. Soft contact involves gently bringing the wafer and mask into contact without additional force. Soft lithography is a popular set of techniques to fabricate microstructures by using elastomeric molds. These techniques can be used in a host of applications, such as biosensors [14]. Hard contact also applies external force to enhance contact, with adjustable force levels.

Vacuum contact improves upon hard contact by using a vacuum to remove air gaps between the wafer and mask for the highest resolution. In cases with brittle substrates, a low vacuum mode minimizes damage by reducing vacuum force. Proximity exposure, a non-contact method, maintains a set distance between the mask and wafer, preventing damage and allowing effective exposure, while incorporating WEC for precision. Resolution increases from proximity to vacuum contact as closer contact between wafer and mask reduces diffractive effects that lower resolution.

5.2.2 Maskless Techniques

The traditional mask-based lithography process is fraught with challenges including high costs and prolonged production delays. According to Menon and Smith [15], the expense of creating a mask set for chip production can surpass $2 million, often for only a limited number of wafers. Additionally, the complexity and financial burden of designing and repairing masks can lead to a conservative approach among chip designers. The resolution in mask-based lithography is highly dependent on the type of mask used—binary masks, phase-shifting masks, and EUV masks each offer different levels of resolution, with EUV masks enabling the smallest feature sizes but at an even greater cost. However, even with advancements in mask technology, the diffraction effects and process variations inherent to mask-based lithography can limit resolution and scalability.

As an alternative, maskless optical lithography presents a solution to these issues by circumventing many of the inherent difficulties associated with mask-based methods. Maskless photolithography, as the name suggests, does not use a permanent photomask. Instead, it utilizes a direct-writing process and creates the pattern where light is projected onto the photoresist in specific patterns, controlled by digital signals. Techniques such as electron beam, ion beam, or laser writing are employed to directly pattern the photoresist without the need for a physical mask. This approach offers flexibility and is particularly advantageous for applications requiring frequent design changes, as modifications can be made digitally without the need for costly mask replacements.

Maskless photolithography operates on the principle of direct pattern writing, which eliminates the need for physical masks. This technique uses a focused beam of light, electrons, or ions, modulated by digital signals [15] that correspond to CAD (Computer-Aided Design) files. This digital control allows for rapid, precise modifications of the pattern without the time and expense of mask fabrication. Additionally, maskless lithography is particularly beneficial in research and development, where iterative design changes are necessary, and in low-volume production, where the high cost of mask fabrication cannot be justified.

Maskless lithography follows a series of well-defined steps that ensure precise pattern transfer onto a substrate. The process begins with substrate preparation, where a wafer is cleaned and coated with a layer of photoresist. Next, a digitally controlled beam-whether it be an electron beam, ion beam, or laser-is directed onto the photoresist according to a pre-designed CAD pattern. Unlike traditional lithography, where a mask determines the exposure pattern, maskless lithography dynamically modulates the beam to expose the photoresist in real-time, allowing for intricate and precise modifications without requiring physical mask alterations. After exposure, the wafer undergoes a development process in which the soluble regions of the photoresist (for positive resist) or the unexposed regions (for negative resist) are removed, leaving behind a structured pattern. This is followed by etching or material deposition, where the patterned areas serve as a guide for subsequent processing steps such as metal deposition, doping, or layer etching. Finally, the remaining photoresist is stripped, revealing the final patterned structure on the wafer. Since maskless lithography enables direct digital control over the patterning process, it allows for real-time adjustments, defect correction, and on-the-fly design changes, making it an ideal solution for applications requiring high flexibility, rapid prototyping, and the fabrication of complex micro- and nanoscale devices.

The substrate coated with photoresist reacts chemically when exposed to the beam; in positive photoresist, exposed areas dissolve, whereas in negative photoresist, they harden. This selective exposure forms the desired pattern directly on the substrate. The ability to repeat this process for multiple layers without changing physical masks significantly enhances production efficiency and reduces error potential. Furthermore, the high-resolution capabilities of maskless lithography, owing to the fine focus of the beam, are crucial for creating micro- and nanoscale features essential in advanced technological applications. While traditional mask-based methods face resolution limitations due to mask defects and optical diffraction, maskless lithography enables higher precision by dynamically adjusting exposure patterns in real-time. This method's adaptability and precision make it a superior choice for rapid prototyping and manufacturing complex devices, where detail and speed are paramount.

5.2.3 Advances in Laser and Electron-Beam Lithography

Recent advancements in laser and electron-beam lithography have significantly enhanced the capabilities and applications of these technologies in micro- and nanofabrication. These advancements have led to improvements in resolution, speed, and flexibility, which are crucial for the development of increasingly complex and miniaturized devices.

Recent advancements in laser and electron-beam lithography have significantly improved their use in micro- and nanofabrication, focusing on enhancements in resolution, speed, and flexibility essential for developing complex, miniaturized devices. Laser lithography has progressed with the adoption of shorter wavelength lasers like DUV and EUV, which allow for finer features necessary for advanced semiconductors. This is complemented by improved optical systems that enhance pattern quality by providing more precise beam control [16].

Electron beam lithography (EBL) is one of the main technologies for the creation of nanoscale patterns. It has benefited from better electron optics and control systems, enabling faster and more stable patterning over large areas. Innovations in resist materials have also reduced the patterning time for complex designs. The integration of these technologies into streamlined manufacturing processes and the introduction of advanced software and automation have improved precision, reduced costs, and expanded applications in fields like nanotechnology and biotechnology, pushing the boundaries of micro- and nanoscale engineering.

Both laser and electron-beam lithography technologies have seen advances in integration with other fabrication processes. This integration allows for more streamlined manufacturing workflows, where patterning, etching, and other fabrication steps are closely coordinated. Such integration is crucial for industries like semiconductor manufacturing, where time and accuracy are critical.

5.3 Development Processes: Wet and Dry Development

After coating and exposing the photoresist, the next crucial step is development. In this process, a developer solution selectively dissolves either the exposed or unexposed regions of the photoresist, depending on whether it is positive or negative. The main goal is to ensure an accurate pattern transfer in photolithography. This step defines circuit features, directly impacting resolution, line edge roughness (LER), and pattern fidelity [8].

This section explores the two main development processes-wet and dry-along with the chemical properties of developer solutions and their impact on resolution and line edge roughness. As illustrated in Fig. 5.2, the development process differs for positive and negative photoresists, where exposed regions are removed in the case of positive resists, while unexposed regions dissolve for negative resists.

5.3.1 Development Process Types

Two primary development techniques are used in modern lithography:

1. Wet Development uses aqueous alkaline solutions or organic solvents to chemically dissolve the resist
2. Dry Development employs plasma-based or vapor-phase etching to remove the resist with high precision

The choice between these methods depends on:

1. Resolution requirements
2. Resist compatibility
3. Processing constraints

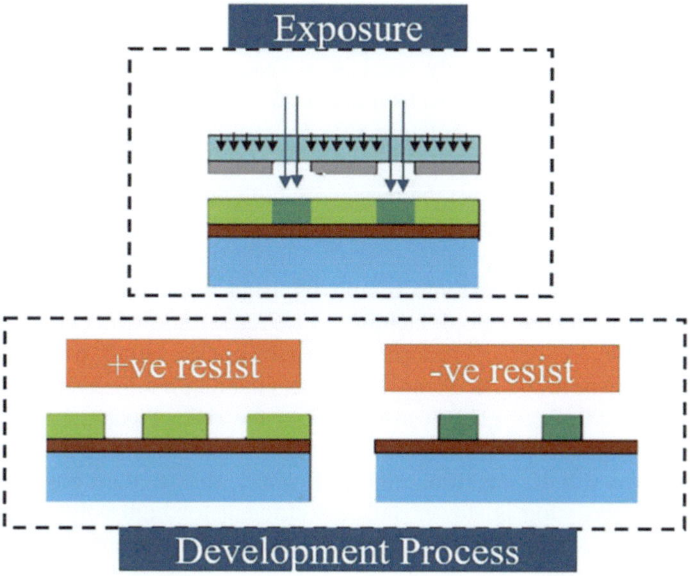

Fig. 5.2 Development process for the silicon wafer

Wet Development Process

Wet development is the most widely used method in conventional lithography, particularly for deep ultraviolet (DUV) and chemically amplified resists (CARs). It involves the dissolution of the photoresist using a liquid developer such as tetramethylammonium hydroxide (TMAH) in positive photoresists, or organic solvents like propylene glycol monomethyl ether acetate (PGMEA) in negative photoresists [17]. The mechanism of wet development follows these key steps:

1. The wafer is submerged in or sprayed with a developer solution.
2. The developer selectively dissolves the exposed or unexposed resist, depending on whether it is a positive or negative photoresist.
3. A rinse step follows, often using deionized water to prevent chemical contamination.
4. The wafer is dried, ensuring that no residual solvent affects pattern definition.

Wet development is widely used in high-volume semiconductor manufacturing due to its high throughput and cost-effectiveness, requiring simpler and less expensive equipment compared to dry processes [18]. However, it has limitations, including lower resolution caused by its isotropic nature, which can lead to undercutting and pattern collapse in sub-10 nm features [19]. Additionally, the process poses a risk of surface contamination, often necessitating extra cleaning steps to ensure wafer integrity.

Dry Development Process

Dry development is increasingly used in advanced lithography, particularly in extreme ultraviolet (EUV) and electron beam lithography (EBL), where wet development may lead to pattern collapse due to capillary forces [20]. Unlike wet development, which relies on liquid-phase dissolution, dry development utilizes plasma-based etching to remove the resist with high precision.

In dry development, the wafer is placed inside a vacuum chamber, where ionized gas species (such as oxygen plasma) bombard the resist surface, breaking molecular bonds and converting the resist into volatile byproducts that are then removed [21]. This method allows for anisotropic etching, where resist removal occurs directionally, minimizing lateral etching and ensuring higher resolution pattern transfer.

Dry development offers excellent resolution, enabling nanoscale patterning with sharp, well-defined edges [21]. It also reduces the risk of pattern collapse since there is no liquid phase, preventing capillary forces from distorting delicate features an advantage for sub-10 nm lithography. However, this method has some downsides. Plasma processing is slower and less efficient for high-volume production, and the need for specialized vacuum and RF power systems increases both cost and operational complexity.

5.3.2 Chemical Nature of Developers

The chemistry of developers influences resolution, pattern fidelity, and defect control. The two primary categories of developers include aqueous alkaline developers and organic solvent-based developers, each optimized for different resist systems and lithographic requirements.

Aqueous Alkaline Developers

Aqueous alkaline developers, especially tetramethylammonium hydroxide (TMAH), are commonly used for positive photoresists due to their strong basicity and compatibility with chemically amplified resists (CARs). In these systems, photoacid generators (PAGs) release acid upon exposure, triggering a solubility change that allows the resist to dissolve in the developer. While TMAH enables high-resolution patterning, its aggressive dissolution can cause rough edges or material loss [8, 17, 20]. Tetramethylammonium hydroxide (TMAH) reacts with a carboxyl (-COOH) functional group in the resist, leading to its dissolution:

$$R - COOH + TMAH \rightarrow (R - COO)^{-} + TMA^{+} + H_2O$$

The chemical reaction involves TMAH deprotonating carboxyl groups, increasing resist solubility. Factors like pH, temperature, and agitation influence dissolution rates. Researchers have explored alternatives like tetrabutylammonium hydroxide (TBAH), which improves resist sensitivity while reducing pattern collapse, making it suitable for EUV and ArFi lithography [17]. Additionally, surfactants in TMAH solutions enhance wetting, improving dissolution efficiency and reducing line-edge roughness, which is crucial for maintaining pattern fidelity in high-resolution lithography.

Organic Solvent-Based Developers

Organic solvent developers are mainly used for negative resists, where exposed regions cross-link and become insoluble. These developers are crucial for molecular glass and metal–organic complex (MOC) resists, which require non-aqueous solvents due to their chemistry. A common choice is methyl isobutyl ketone (MIBK), often mixed with isopropyl alcohol (IPA) to selectively dissolve unexposed areas [8, 20]. The development process involves chemical reactions such as ester hydrolysis in chemically amplified resists (CARs), forming carboxylic acids, or NMP (N-Methyl-2-Pyrrolidone) breaking hydrogen bonds to solvate resist fragments.

$$R - COOCH_3 + H_2O \rightarrow R - COOH + CH_3OH$$

$$R - N - R\prime + NMP \rightarrow Solubleresistfragments$$

However, organic solvent developers pose environmental and safety concerns due to their volatility, prompting research into greener alternatives with improved sustainability profiles.

Impact of Developer Chemistry on Pattern Quality

The chemical composition of a developer significantly influences critical dimension (CD) control, line roughness, and defect reduction. TMAH, while effective, can lead to overdevelopment and pattern collapse if not carefully optimized. Conversely, TBAH has demonstrated improved pattern stability while maintaining high resolution. In negative photoresists, the choice of organic developer impacts the degree of cross-linking, affecting etch resistance and pattern uniformity [17, 20].

Future Trends in Developer Chemistry

As semiconductor manufacturing pushes beyond the sub-5 nm scale, developers must keep up to ensure precision and minimize defects. One key focus is on metal-ion-free alkaline developers to prevent contamination in EUV lithography [22]. At the same time, researchers are exploring non-toxic organic solvents to improve sustainability without sacrificing performance [23]. Hybrid aqueous-organic developers are emerging as a promising option, offering better dissolution control for next-generation resists. Additionally, surfactant-enhanced TMAH and TBAH developers are helping to improve resist adhesion and reduce defects. Another major challenge is tackling stochastic effects which are random variations in pattern formation with future formulations aiming for better diffusion control and pattern stability. These advancements will be crucial for maintaining the pace of semiconductor scaling and ensuring long-term reliability.

5.3.3 *Impact on Resolution and Line Edge Roughness (LER)*

In modern semiconductor fabrication, pattern resolution and line edge roughness (LER) are critical to device performance. Resolution defines the smallest reliably patterned feature, while LER and line width roughness (LWR) refer to edge variations that can introduce defects and signal delays, especially in sub-10 nm nodes. These roughness effects can be caused by chemical inhomogeneities (Uneven distribution of photoacid generators (PAGs) in chemically amplified resists leads to localized variations in dissolution rates), diffusion dynamics, and resist-developer interactions. The choice between wet and dry development significantly impacts resolution and LER, as factors like dissolution behavior, chemical selectivity, etch precision determine how accurately patterns transfer onto the wafer [8, 19].

Impact of Wet Development on Resolution and LER

Wet development poses resolution limitations due to its isotropic etching profile, which can lead to undercutting and non-uniform dissolution. Additionally, capillary forces increase the risk of pattern collapse in high-aspect-ratio structures. Studies show that TMAH-based developers achieve resolutions up to 40 nm, but further scaling requires dry development techniques. Wet developers also contribute to LER, particularly when overexposure leads to excess dissolution. Surfactants and buffering agents help control LER by stabilizing chemical gradients, while proper rinsing with deionized water prevents residue buildup and roughness [17, 19].

Impact of Dry Development on Resolution and LER

Dry development offers superior resolution through anisotropic etching, ensuring precise feature definition with minimal lateral spread. Plasma-based resist removal enables sub-10 nm patterning, making it essential for EUV and EBL lithography [24]. It also reduces LER by using directional etching, which removes resist material vertically and prevents sidewall degradation. However, excessive plasma power can induce microscopic surface roughness, necessitating careful gas composition control (O_2, CF_4, CHF_3 mixtures) to minimize ion damage and maintain smooth edges [25].

Minimization Strategies for LER and Resolution Enhancement

To minimize LER and enhance resolution, several strategies are employed. Optimized developer formulations with buffering agents regulate dissolution in wet processes, while post-development annealing smooths rough edges. Advanced rinse techniques, like supercritical CO_2 drying, prevent capillary-induced roughness. In dry development, precise plasma parameter control adjusting RF power, pressure, and gas composition reduces ion bombardment defects. These methods collectively improve pattern accuracy and maintain feature integrity.

Computational Lithography and Advanced Patterning Techniques

The semiconductor industry has been subjected to continuous progression from 7 nm; 5 nm and now it is on its way to sub 3 nm regime. The ever-increasing sophistication demands posed by these scaled down processes cannot be catered by the traditional photolithographic processes. The issues like optical diffraction, process variability and resist material constraints are the major constraints for the existing solutions to be an ideal fit for smaller nodes. To address these challenges computational lithography; mathematical modelling techniques and machine learning algorithms are being employed to optimize and improve the pattern resolution. Some advanced techniques, like Inverse lithography technology; source mask optimization and optical proximity correction are being used to compensate for the optical distortions at the subwavelength patterning.

5.4 Post-Exposure and Post-Development Treatments

The method employed to reduce the standing wave phenomenon is post-exposure bake (PEB). The precise mechanism remains contentious; nevertheless, it is generally accepted that elevated temperatures (100 °C-130°C) promote the diffusion of the photoactive chemical, so successfully mitigating standing wave ridges. The adverse effects of elevated temperatures on the photoresist, previously addressed with prebaking, similarly pertain to post-exposure baking (PEB), necessitating meticulous optimization of baking settings. Furthermore, research demonstrates that the diffusion rate of the photoactive molecule is affected by prebake conditions, with findings indicating that the presence of solvent facilitates diffusion during post-exposure bake (PEB). Thus, a reduced prebake temperature results in enhanced diffusion at a certain PEB temperature.

5.4.1 Prebake (Softbake)

The complete process for UV lithography process steps which is a critical technique in the fabrication of microfluidic components, involving the exposure of a spin-coated photoresist layer to UV light through a shadow mask is shown in Fig. 5.3. During this process, the exposed regions undergo crosslinking, while the unexposed, non-crosslinked portions are dissolved and removed during the development stage, leaving only the desired structures on the wafer. The soft-bake process is a critical step in photolithography, primarily used to remove solvents from the photoresist layer applied to the substrate. This ensures proper adhesion and uniformity of the resist film. Typically, the softbake is performed at a temperature range of 90 °C to 120 °C for a duration of 60 to 120 s, depending on the specific photoresist material used. The process can be conducted using either a hot plate or an oven, with the hot

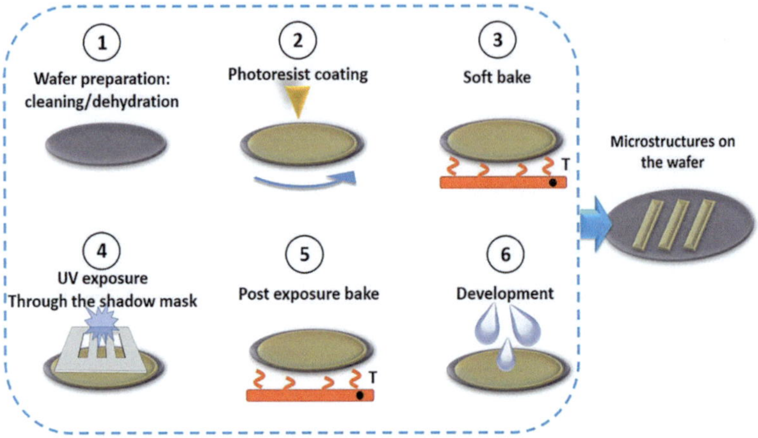

Fig. 5.3 Process diagram for UV lithography of silicon wafer [27]

plate being preferred for its rapid heating capabilities, which helps in achieving a more uniform temperature distribution across the substrate. It is essential to control the softbake conditions carefully, as excessive temperature or time can lead to resist degradation, while insufficient conditions may result in incomplete solvent removal. After the soft bake, the photoresist is ready for exposure to UV light, which is crucial for defining patterns on the substrate [26].

5.4.2 Post-Development Bake (Hard Bake)

The hard bake, also known as the post-development bake, is performed after the development of the photoresist to enhance its thermal stability and improve adhesion to the substrate. This step is crucial for ensuring the durability of the resist pattern during subsequent processing steps. Hard bake temperatures are set near the resin's glass transition temperature (80–120 °C for diazo naphthoquinone (DNQ)/novolac resists) but above the sensitizer's decomposition point, as photosensitivity is no longer needed. Baking above 110 °C decomposes DNQ, enabling cross-linking with novolac resin. At 120–130 °C, hardening competes with resist flow, causing deformation and rounded edges, with larger features being more prone to thermal-induced distortion [28]. This process can be conducted in a convection oven or on a hot plate, with the oven providing more uniform heating across the substrate. It is important to ensure that the temperature is evenly distributed to avoid any thermal gradients that could affect the resist. During the hard bake, the photoresist undergoes cross-linking, which increases its chemical resistance and mechanical strength, making it less susceptible to damage during etching or other processing steps. After the hard bake, the substrate is typically cooled down to room temperature before proceeding to the next fabrication steps, ensuring that the resist pattern remains intact and stable [29].

5.4.3 Chemical and Thermal Curing

Chemical curing enhances photoresist stability by promoting cross-linking, improving resistance to solvents and etchants, while thermal curing, performed at 120–180 °C during hard bake, strengthens adhesion and mechanical properties. Combining both methods enhances durability for etching and ion implantation. Innovations like ammonia–nitrogen baking gases reduce shrinkage and critical dimension variations, while advanced photo initiators refine patterning, highlighting the ongoing advancements in photoresist optimization for semiconductor fabrication.

Impact of Chemical and Thermal Curing on Adhesion and Resolution Retention

Chemical curing enhances the cross-linking density within the photoresist, which significantly improves adhesion to the substrate. This increased adhesion is crucial for maintaining the integrity of the resist pattern during subsequent processing steps, such as etching and ion implantation. Thermal curing, performed during the hard bake, further promotes cross-linking through elevated temperatures, leading to a more robust resist film. This process not only strengthens the adhesion but also contributes to the overall mechanical stability of the resist pattern. The combination of both curing methods results in improved resolution retention, as the enhanced chemical and thermal properties of the resist allow for finer features to be maintained without distortion during processing. This is particularly important in advanced semiconductor fabrication, where precision is paramount. Ultimately, the synergistic effect of chemical and thermal curing ensures that the photoresist can withstand the rigors of fabrication processes while preserving the desired pattern fidelity and adhesion characteristics.

5.4.4 Emerging Techniques for Post-Processing

UV Exposure for Enhanced Resolution

Utilizing UV light in post-processing can significantly improve the resolution of patterns in photoresist materials. This technique allows for the fine-tuning of the resist profile, enabling the creation of smaller features that are essential for advanced semiconductor applications. The precision of UV exposure helps in achieving high fidelity in the final patterns. UV curing is used to harden polymeric resists like novolac, poly (hydroxy styrene), and polyacrylates, allowing higher post bake temperatures without damaging the pattern. However, the high optical absorbance of aromatic resins below 300 nm limits cross-linking efficiency. Heating above 150 °C enhances deep UV (DUV) curing, forming a thermally stabilized surface crust. At even higher temperatures, oxygen-mediated radical cross-linking strengthens the resist, enabling it to withstand thermal processes up to 210 °C with minimal deformation. Additionally, multifunctional DNQ/novolac resins undergo DUV-induced ester-based cross-linking [30].

Plasma Treatment

Emerging techniques also include plasma treatment, which can modify the surface properties of the photoresist. This method can enhance adhesion and improve the resist's resistance to etching, thereby contributing to better pattern retention and overall performance in microfabrication processes.

Electron Beam Curing

Electron beam curing relies on inelastic collisions between high-energy electrons and polymeric resist molecules. As the electrons pass through the resist film, they lose energy and ionize the resist molecules before being absorbed. Secondary electrons, moving at slower speeds, also generate ions along the primary electron trajectory. Within a picosecond of ionization, molecular rearrangement occurs, leading to thermal relaxation or the breaking of valence bonds. The resulting radicals either recombine or form new bonds with adjacent polymer chains, promoting cross-linking. A key mechanism involves hydrogen abstraction from methylene groups, generating reactive radicals that rapidly bond with others, transforming the linear polymer into a three-dimensional cross-linked network. This process continues with further electron beam exposure until full cross-linking of the resist film is achieved [31, 32].

5.4.5 Enhancement of Durability & Adhesion of Photoresist Through Post-Processing Techniques

In addition to optimizing the softbake and hard bake processes, advancements in photolithography are increasingly focusing on the integration of real-time metrology systems that can provide immediate feedback during each stage of fabrication. These systems enable manufacturers to monitor not only temperature profiles but also CD variations as they occur, thereby facilitating rapid adjustments to processing parameters. For instance, utilizing a continuous wave CO_2 laser for post-exposure baking has shown potential to significantly reduce acid diffusion rates compared to conventional methods, which could lead to improved pattern fidelity and surface roughness while minimizing dose requirements [33]. Furthermore, incorporating machine learning algorithms into this monitoring process may enhance predictive capabilities, allowing for proactive modifications to resist formulations or exposure conditions based on historical data trends, ultimately pushing the boundaries of resolution achievable in modern semiconductor devices.

5.4.6 Post-Processing Techniques in IC Fabrication

Integrated circuits may require baking prior to utilization on any board type, whether prototype or production. Integrated circuits are typically stored in an electrostatic discharge-safe and moisture-sealed container upon leaving the manufacturing. Even when procured from a respected distributor, they should be packaged in this manner. Typically, they contain a moisture indicator and a desiccant. The necessity for bake-out is contingent upon the Moisture Sensitivity Level (MSL) of the component and the moisture exposure it has experienced. Certain components require compulsory bake-out [34]. An effective Assembly House will assess the MSL of the components and perform the necessary bake-out procedures accordingly. If you have not stored your components according to the manufacturer's specifications, you should consistently get them baked out. Moisture can lead to various complications: it may result in the formation of bubbles beneath plastic encapsulation (bubbling), induce stress on the substrate (potentially causing premature failure and/or non-compliance with specifications), and alter the impedance of the core and pre-preg in PCBs for impedance-controlled routes (occasionally necessitating the baking of PCBs as well).

References

1. M.D. Tyona, Adv. Mater. Res. **2**, 195–208 (2013)
2. O. Büyükalaca, H. Bulut, Appl. Energy **77**, 187–204 (2004)
3. L.F. Hakim, D.M. King, Y. Zhou, C.J. Gump, S.M. George, A.W. Weimer, Adv. Funct. Mater. **17**, 3175–3181 (2007)
4. D.E. Weidner, S. Mahmoodi, *Physics of Fluids* 35, (2023)
5. Z. Weng, X. Huang, S. Peng, L. Zheng, L. Wu, *Nat. Commun.* 14, (2023)
6. W. Khuen Ho, A. Tay, L. Lay Lee, C. D. Schaper, *Control Eng. Pract.* 12, 881–892 (2004)
7. X. Dong, X. Pan, X. Gao, H. Fang, *Coatings* 10, (2020)
8. Y. Zhang, H. Yu, L. Wang, X. Wu, J. He, W. Huang, C. Ouyang, D. Chen, B.E. Keshta, *Adv. Colloid Interface Sci.* 329, (2024)
9. J. Park, K. Shin, C. Lee, Int. J. Precis. Eng. Manuf. **17**, 537–550 (2016)
10. M.A. Shah, D.-G. Lee, B.-Y. Lee, S. Hur, IEEE Access **9**, 140079–140102 (2021)
11. A.E. Adeoye, O.A. Adeaga, K. Ukoba, *Nigerian Journal of Technology (NIJOTECH)* 43, 479–489
12. L.S. Melvin, Adv. Opt. Technol. **10**, 85–86 (2021)
13. M.M. Ghobashy, S.A. Alkhursani, H.A. Alqahtani, T.K. El-damhougy, M. Madani, *Materials Science and Engineering: B* 301, (2024)
14. A. Mujahid, N. Iqbal, A. Afzal, Biotechnol. Adv. **31**, 1435–1447 (2013)
15. R. Menon, A. Patel, D. Gil, H.I. Smith, Mater. Today **8**, 26–33 (2005)
16. W. Zhou, D. Bridges, R. Li, S. Bai, Y. Ma, T. Hou, A. Hu, Researchgate **5**, 228 (2016)
17. M. Harumoto, J.J. Santillan, C. Nakayama, Y. Tanaka, T. Motono, M. Asai, T. Itani, J. Photopolym. Sci. Technol. **32**, 321–326 (2019)
18. M.O.d. Beeck, N. Samarakone, K.H. Baik, L.V.d. hove, D. Ritchie, *Advances in Resist Technology and Processing VII* 1262, 70–80 (1990)
19. S. Mori, S. Shirayone, S. Irie, N. Matsuzawa, H. Oizumi, E. Yano, S. Okazaki, A. Miyafuji, T. Watanabe, H. Kinoshita, T. Oshio, K. Sugisaki, Microelectron. Eng. **53**, 689–692 (2000)
20. J. Cen, Z. Deng, S. Liu, Polym. Chem. **15**, 4599–4614 (2024)

21. D.V. Nicolau, P. Jansz, G. Metcalfe, G. Wild, S. Hinckley, A comparison of wet and dry etching to fabricate a micro-photonic structure for use in OCT, Biomedical Applications of Micro- and Nanoengineering IV and Complex Systems, p. 727016, (2008)
22. W.R.B.a.A.R.N. Cynthia Zee, *Advances in Resist Technology and Processing* **920**, 154–161 (1988)
23. X. Shao, A. Guerrero, Y. Gu, *SEMICON China 2004 SEMI Technology Symposium* 1–9 (2004)
24. D. Baek, S.H. Lee, B.-H. Jun, S.H. Lee, pp. 217–233, (2021)
25. Y. Sha, S. Yao, M. Jiang, H. Yang, D. Liang, C. Wang, F. Wang, E. Tian, J. Xi, Y. Jiang, J. Shi, M. Burkhardt, C. van Lare, Photoresist process simulation to study line edge roughness, Optical and EUV Nanolithography XXXVII, (2024)
26. J. Wiley, S. Ltd, *John Wiley & Sons.* (2008)
27. G. Konoplev, D. Agafonova, L. Bakhchova, N. Mukhin, M. Kurachkina, M.-P. Schmidt, N. Verlov, A. Sidorov, A. Oseev, O. Stepanova, A. Kozyrev, A. Dmitriev, S. Hirsch, *Biomedicines* 10, (2022)
28. U. Okoroanyanwu, *Micro Nano Eng.* 24, (2024)
29. D. Steele, A. Coniglio, C. Tang, B. Singh, Metrology. Inspection, and Process Control for Microlithography XVI **4689**, 517–530 (2002)
30. M.J. Leeson, W. Yueh, P.I. Tattersall, A. Pawloski, S.M. Grayson, C.G. Willson, Chem. Mater. **16**, 1763–1769 (2004)
31. J. Raghavan, M.R. Baillie, Polym. Compos. **21**, 619–629 (2004)
32. W.R. Livesay, A.L. Rubiales, M. Ross, S. Woods, S. Campbell, Advances in Resist Technology and Processing X **1925**, 426–436 (1993)
33. B. Jung, C.K. Obera, M.O. Thompsona, T.R. Younkinb, M. Chandhokb, Addressing challenges in lithography using sub-millisecond post exposure bake of chemically amplified resists, Adv. Resist. Mater. Process. Technol. XXVIII, pp. 382–387, (2011)
34. X. Ma, K.M.B. Jansen, G.Q. Zhang, W.D. van Driel, O. van der Sluis, L.J. Ernst, C. Regards, C. Gautier, H. Frémont, Microelectron. Reliab. **50**, 1654–1660 (2010)

Chapter 6
Photoresist Applications in Microsystems

The rapid advancement of microsystems technology has transformed industries ranging from healthcare and telecommunications to energy and aerospace. At the core of this progress is photolithography, a process enabled by photoresists: light-sensitive materials that allow for the precise patterning of microscale and nanoscale structures. Initially developed for semiconductor manufacturing, photoresists are now essential in the production of micro-electro-mechanical systems (MEMS), biomedical devices, optical components, and energy storage solutions. Their ability to define complex geometries, adapt to different materials, and integrate with new fabrication techniques makes them a cornerstone of modern microsystems engineering.

Microsystems, such as MEMS accelerometers, lab-on-a-chip platforms, and microfabricated batteries, require precision, miniaturization, and efficient material integration. Photoresists help achieve these goals by enabling high-resolution lithography, 3D structuring, and multi-material compatibility. For example, in MEMS sensors and actuators, photoresists like SU-8 and SPR 220–7 allow for the fabrication of tiny mechanical components such as proof masses, comb drives, and suspension springs. These structures are crucial for devices like gyroscopes and pressure sensors, where accuracy and reliability are essential [1, 2]. In biomedical applications, photoresists play a key role in developing lab-on-a-chip devices, which perform diagnostics and drug testing using extremely small liquid volumes [3, 4].

Photoresists also enable advancements in optomechanical microsystems, which integrate optical and mechanical components for applications such as high-resolution imaging and photonic communication. Techniques like thermal reflow and grayscale lithography use photoresists to create microlenses and optical waveguides, improving sensor performance and light manipulation [5]. Meanwhile, in energy storage and conversion, photoresist-based patterning helps optimize electrode architecture, electrolyte distribution, and microfluidic pathways. These improvements directly enhance the energy density and efficiency of microbatteries and fuel cells [6, 7].

Recent advancements in photoresist materials are expanding their capabilities even further. Hydrogen silsesquioxane (HSQ), a highly stable inorganic resist,

K. Wang, *Photoresist Technology in Microsystems: Principles, Processes and Applications*, Springer Series in Advanced Microelectronics 15,
https://doi.org/10.1007/978-981-95-1606-3_6

enables the fabrication of sub-100 nm photonic crystals and optical MEMS. Meanwhile, off-stoichiometry thiol-ene (OSTE) polymers offer customizable mechanical properties, making them ideal for organ-on-chip systems that mimic human tissues for biomedical research [8]. These materials help address longstanding challenges related to scalability, durability, and performance in microfabrication.

This chapter explores how photoresists contribute to the development of MEMS sensors, biomedical devices, optical systems, and energy storage technologies. It also examines the impact of new materials like HSQ and OSTE in specialized applications. By analysing fabrication methods, material innovations, and real-world case studies, this chapter highlights how photoresists continue to drive the future of microsystems technology, enabling smaller, smarter, and more efficient devices.

6.1 MEMS Devices: Sensors and Actuators

Photoresist contributions to accelerometers and gyroscopes: The integration of photoresists into MEMS has revolutionized the fabrication of sensors and actuators, enabling unprecedented precision, scalability, and compatibility with semiconductor manufacturing processes. This section focuses on the critical role of photoresists in the development of MEMS accelerometers and gyroscopes, highlighting their contributions to structural patterning, material compatibility, and functional performance. By examining fabrication workflows, material innovations, and emerging techniques such as 3D printing, this analysis underscores how photoresists have become indispensable in advancing MEMS technology.

6.1.1 Photoresist-Driven Fabrication of MEMS Accelerometers

MEMS accelerometers rely on photoresists to define microscale mechanical structures, such as proof masses and suspension springs, which are critical for detecting inertial forces. The fabrication process typically begins with silicon wafer preparation, where photoresists enable photolithography to transfer geometric patterns onto the substrate. For instance, in capacitive accelerometers, a sequence of oxidation, photolithography, and etching steps creates V-grooves that form the basis of the sensing element. During photolithography, a positive photoresist is applied to the silicon dioxide layer, exposed to UV light through a patterned mask, and developed using tetramethyl ammonium hydroxide (TMAH) to reveal the underlying oxide for etching. This process ensures precise alignment of the proof mass and interdigitated electrodes, which are essential for measuring capacitance changes under acceleration [1, 9].

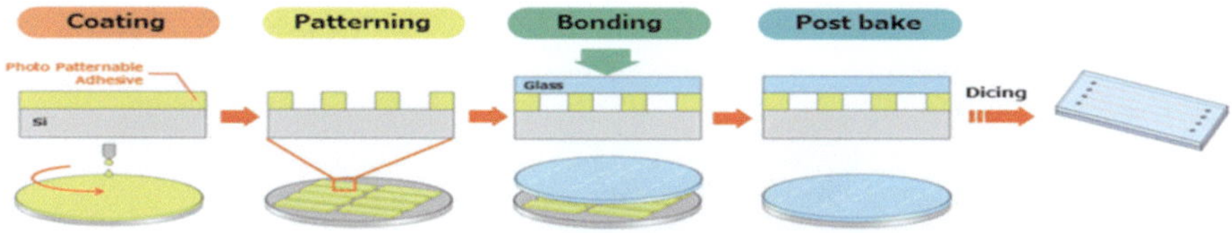

Fig. 6.1 Process of usage of photoresist for MEMS Cantilever based devices

Recent advancements in accelerometer design leverage novel photoresist-based techniques to enhance performance. For example, glass-silicon composite wafers utilize photoresist masks during deep reactive ion etching (DRIE) to create high-aspect-ratio structures with minimal sidewall roughness. Additionally, anodic bonding of glass and silicon substrates a process facilitated by photoresist-patterned electrodes; ensures hermetic packaging, which is vital for maintaining sensor stability in harsh environments. The emergence of 3D-printed MEMS accelerometers further demonstrates photoresist versatility, where two-photon polymerization of IP-S resin constructs intricate geometries, such as T-shaped cantilevers and proof masses, in a single manufacturing step. This approach eliminates the need for multi-mask lithography, reducing fabrication complexity while enabling customization for niche applications [1, 2] (Fig. 6.1).

6.1.2 Photoresist Applications in MEMS Gyroscopes

MEMS gyroscopes, which measure angular velocity via Coriolis force detection, depend on photoresists to fabricate vibrating elements and sensing electrodes. The working principle involves energy transfer between two vibrational modes, a phenomenon enabled by photolithographically defined comb drives and suspension springs. A key challenge lies in achieving symmetrical resonator structures to minimize quadrature error, which photoresists address through high-resolution patterning. For instance, anisotropic potassium hydroxide (KOH) etching of silicon, guided by photoresist masks, produces trapezoidal cross-sections that enhance mechanical stability. Subsequent metallization steps employ shadow masks; a photoresist-free alternative to deposit aluminium leads on high-step-height surfaces, avoiding coating inconsistencies associated with traditional lift-off processes [1, 10].

Piezoelectric gyroscopes represent another photoresist-driven innovation, where sol–gel-deposited lead zirconate titanate (PZT) layers are patterned using photoresist masks to create stress-sensitive elements. These layers, when combined with photolithographically defined aluminium interconnects, enable precise detection of capacitance changes caused by Coriolis-induced displacements. However, the brittleness of polysilicon and the complexity of multi-level masking remain challenges, necessitating advanced photoresist materials with improved etch selectivity and thermal stability [1, 9] (Fig. 6.2, Table 6.1).

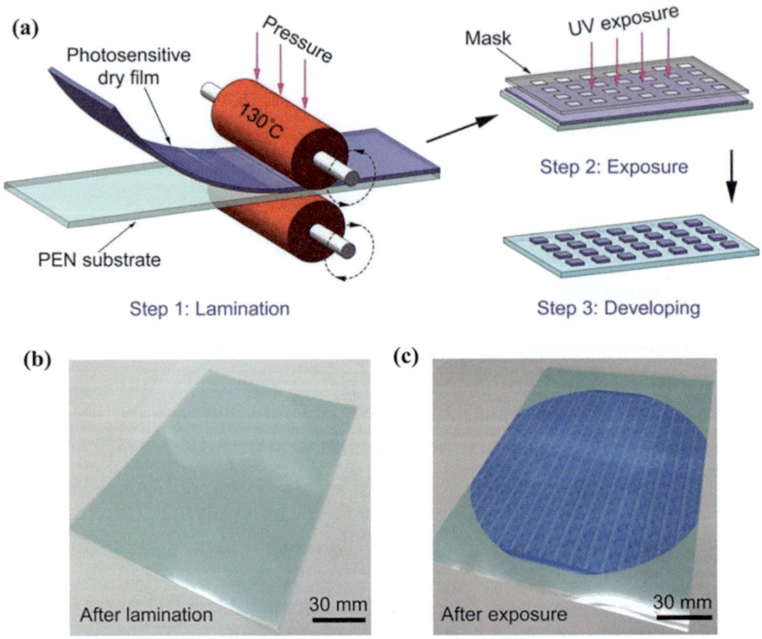

Fig. 6.2 Typical dry film photoresist lamination process demonstrating **a** Lamination of a DFP film on a substrate using roll on technique **b** The substrate after lamination and **c** After exposure with UV light showing the patterns [11]

Table 6.1 Photoresist applications in MEMS Gyroscopes [1, 2, 9, 10]

Property	SU-8	SPR 220–7	Ordyl P-50100	AZ 9260
Maximum aspect ratio	> 20:1	5:01	5:01	10:01
Thickness range (μm)	1–500	5–100	50–100	May-30
Young's modulus (GPA)	4.02	3.5	2.3	2.9
Glass transition temperature (°c)	> 200	115	130	110
Residual stress (MPa)	16–19	28	12	32
Resolution (μm)	2	0.5	10	0.8
Chemical resistance	Excellent	Good	Moderate	Good
UV sensitivity wavelength (nm)	365	365–436	350–400	365–436

6.1.3 Comparative Analysis of Photoresist Techniques

The selection of photoresists for MEMS fabrication hinges on their mechanical, optical, and processing characteristics. SU-8, SPR 220–7, and Ordyl P-50100 represent three distinct material systems that address diverse MEMS requirements.

SU-8, an epoxy-based negative-tone photoresist, dominates applications requiring ultra-thick (>500 μm) layers and high aspect ratios (>20:1). Its low optical absorption at 365 nm enables uniform crosslinking through thick films, making it ideal for inertial sensors with deep suspension springs. For example, SU-8's Young's modulus (~4.02 GPa) and thermal stability (T_g > 200 °C) allow it to serve as both a structural material and a mold for electroplated nickel components in accelerometers [2]. However, SU-8's brittleness and residual stress (16 to 19 MPa on silicon) limit its use in flexible MEMS components.

In contrast, SPR 220–7, a novolak-based positive photoresist, excels in high-resolution patterning for metallization steps. With a typical thickness range of 1 to 100 μm, SPR 220–7 achieves submicron resolution through i-line (365 nm) exposure, enabling precise definition of capacitive comb fingers in gyroscopes. Its chemical resistance to TMAH etchants permits simultaneous patterning of aluminum interconnects and silicon microstructures. However, SPR 220–7's limited thickness capability and solvent sensitivity restrict its use in multilayer MEMS fabrication [1, 9].

Ordyl P-50100, a dry-film photoresist, addresses non-planar substrate challenges in 3D MEMS. As a solvent-free laminate, it adheres to high-topography surfaces; critical for packaging MEMS accelerometers with integrated ASICs. Ordyl's 50–100 μm thickness range and rapid processing (no spin-coating) make it suitable for industrial-scale production of pressure sensors. However, its lower aspect ratio (~5:1) compared to SU-8 limits feature miniaturization [2] (Table 6.2).

6.1.4 Hybrid and Molecular Photoresists

Recent advancements in non-chemically amplified molecular resists (n-CARs) address limitations of traditional polymer resists, particularly for nodes smaller than 20 nm. Molecular resists like nitrobenzene derivatives reduce line-edge roughness (LER) by eliminating polymer chain entanglement, achieving feature sizes of 3–5 nm [1]. These materials are critical for extreme ultraviolet (EUV) lithography, where high absorbance and thin resist films are essential due to the low photon count of EUV sources14. Inorganic–organic hybrid photoresists further enhance EUV compatibility by incorporating high-atomic-number metals (e.g., tin, hafnium) that improve photon absorption and etch resistance. For instance, metal-oxo clusters functionalized with organic ligands enable sub-10 nm patterning with aspect ratios exceeding 10:1, making them ideal for nanoelectromechanical systems (NEMS) [1, 10].

The interplay between photoresist selection and MEMS performance is exemplified in automotive gyroscopes. SU-8's optical transparency (>90% at 600 nm) enables integration of waveguide-based optical readout systems, reducing quadrature error by 40% compared to capacitive designs. Meanwhile, SPR 220–7 facilitates the fabrication of aluminum-encapsulated PZT actuators, achieving 0.1° bias stability over −40 °C to 125 °C [1, 9].

Table 6.2 Comprehensive photoresist comparison [1, 2, 9, 10]

	SU-8 (Negative)	SPR 220–7 (Positive)	Ordyl P-50100 (Negative)	AZ 9260 (Positive)	ma-N 1400 (Negative)	IP-Dip (Negative)
Chemical Composition	Epoxy-based	Novolak/DNQ-based	Acrylic polymer	Novolak/ DNQ-based	Phenolic resin	Acrylate-based
Maximum Thickness (Î¼m)	500 +	100	100	30	2	150 (3D printing)
Minimum Feature Size (Î¼m)	2	0.5	10	0.8	0.1	0.2
Aspect Ratio	> 20:1	5:01	5:01	10:01	3:01	25:01:00
Young's Modulus (GPa)	4.02	3.5	2.3	2.9	9.7	4.6
Thermal Stability (Â°C)	> 200	115	130	110	160	120
Residual Stress (MPa)	16–19	28	12	32	45	29
Optical Transparency	High (>90% at 600 nm)	Moderate	Low	Low	Low	High
Biocompatibility	Good	Poor	Moderate	Poor	Poor	Excellent
Process Complexity	Moderate	Low	Very Low	Low	High	High
Developer	PGMEA	TMAH	Sodium carbonate	TMAH	TMAH	Propylene glycol methyl ether
Primary MEMS Applications	High-AR structures, micro-fluidics	Fine electrodes, standard patterning	Packaging, lamination	High-current electroplating	Nanoimprint masters	3D micro-fabrication

In bioMEMS, SU-8's biocompatibility supports implantable accelerometers for gait analysis. Surface modification with poly-L-lysine coatings enhances cell adhesion on SU-8 cantilevers, enabling real-time monitoring of cardiomyocyte contractions. Conversely, Ordyl P-50100 enables rapid prototyping of disposable microfluidic cartridges, with burst pressures exceeding 15 psi in hybrid PDMS/SU-8 valves [2].

Latent image simulations using tools like THB-151N optimize exposure parameters for high-aspect-ratio structures. For example, polychromatic light attenuation models predict cross-linked columnar features in 100 μm-thick resists, minimizing development time by 30% while maintaining $\pm$ 1% dimensional accuracy [9].

The shift from SPR 220–7 to SU-8 in the 2000s marked a paradigm shift toward monolithic MEMS. Early airbag accelerometers relied on SPR 220–7 for aluminum etching masks, but SU-8's introduction enabled direct fabrication of 300 μm-deep inertial masses, reducing package sizes by 60%. The 2010s saw Ordyl P-50100 bridge the gap between MEMS and PCB manufacturing, allowing wafer-level bonding at costs 70% lower than anodic methods [2].

Despite their widespread adoption, photoresists face limitations in MEMS applications. Stiction, the adhesion of microstructures during wet etching, remains a persistent issue, particularly for devices with high surface-area-to-volume ratios. Advanced anti-stiction coatings, such as fluorinated self-assembled monolayers (SAMs), are being integrated into photoresist formulations to mitigate this problem [10]. Additionally, residual stress in cured photoresist films can induce warping in thin diaphragms, necessitating stress-optimized materials for optical MEMS components [1, 9].

Future research focuses on stimuli-responsive photoresists that enable reconfigurable MEMS architectures. For instance, temperature-sensitive azobenzene-containing resists allow post-fabrication tuning of grating periods in MEMS spectrometers. Similarly, pH-responsive hydrogels are being explored for bioMEMS applications, where environmental triggers can modulate device functionality [1, 10].

6.2 BioMEMS and Microfluidics: Lab-On-A-Chip Devices

A lab-on-a-chip is a small-scale device capable of performing meaningful operations in various fields, such as medicine and biology, and delivering results that substitute large-scale laboratory work, which typically requires more time and expense. Research on such devices focuses on applications including human diagnostics, DNA analysis, and, to a lesser extent, chemical synthesis. As a diagnostic tool, it can be used for one-time applications, such as COVID-19 test kits [12], or for multiple or reusable applications in toxicology and drug testing [13, 14].

Lab-on-a-chip devices not only demonstrate the ability to integrate into systems but also show good performance compared to large-scale technologies. They use small amounts of liquid and do not require huge quantities of materials for production.

Microfluidics, the science of manipulating picolitres of fluids, is a key application in manufacturing microminiaturized devices and serves as the core technology behind lab-on-a-chip systems. It can integrate a vast number of microchannels, each measuring mere micrometres, onto a single chip that fits in the palm of your hand. This process enables efficient handling of small fluid volumes [15].

Microfluidic devices require specific photoresists to accurately fabricate intricate micro-scale channels and structures essential for their function. Photoresists are light-sensitive materials that, upon exposure to ultraviolet (UV) light, undergo chemical changes, enabling selective patterning on substrates. The precision and performance of microfluidic devices heavily depend on the properties of the chosen photoresist.

One of the main points to consider is high resolution and aspect ratios. Microfluidic applications demand the creation of fine features with high aspect ratios to ensure precise control over fluid flow. Photoresists must facilitate the fabrication of these detailed patterns without compromising structural integrity. For instance, SU-8, an epoxy-based negative photoresist, is renowned for producing high-aspect-ratio structures with near-vertical sidewalls, making it ideal for microfluidic device fabrication [16].

Chemical and thermal stability are also essential factors. Devices often operate under various chemical environments and temperatures, which is why the photoresist material must withstand these conditions without degrading. SU-8 exhibits excellent chemical resistance and thermal stability, ensuring the durability of microfluidic structures during prolonged use [17].

Biocompatibility is another critical consideration, especially in applications involving biological samples, such as lab-on-a-chip devices. The materials used must be biocompatible to prevent adverse reactions with biological entities. SU-8 photoresist has been demonstrated to possess high biocompatibility, making it suitable for bio-MEMS and other biomedical applications [16].

As one of the most popular and previously mentioned photoresists, SU-8 is a widely used epoxy-based negative photoresist (Fig. 6.3). It is preferred for its ability to create thick, high-aspect-ratio structures with excellent resolution. Its chemical stability and biocompatibility further make it a standard choice in microfluidic device fabrication [15].

Off-Stoichiometry Thiol-Ene (OSTE) polymers provide a versatile platform for microfluidic applications due to their tunable mechanical properties and surface chemistries. They enable rapid prototyping and facilitate strong, biocompatible bonding, making them highly advantageous for lab-on-a-chip devices [18].

Dry-Film Photoresists are valued for their excellent flatness and uniformity, which are critical for fabricating precise microfluidic chip moulds. These films can be laminated to achieve the desired thickness and are compatible with standard photolithography techniques, offering a streamlined and efficient fabrication process [19].

BioMEMS (Biological Microelectromechanical Systems) require specific photoresists to ensure the precise fabrication of microscale structures necessary for their functionality. These devices are used in medical diagnostics, drug delivery, and tissue engineering, where the properties of photoresists play a crucial role in device

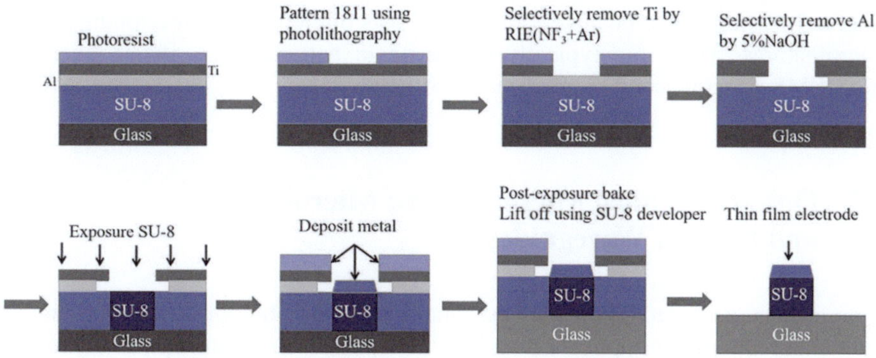

Fig. 6.3 Procedure for fabricating a self-aligned metal thin film on an SU-8 pedestal

performance. The key reasons BioMEMS require specific photoresists similar to lab-on-a-chip technology include high resolution and aspect ratios, chemical and thermal stability, biocompatibility, and mechanical strength and durability.

First, these structures demand the fabrication of microstructures with high resolution and high aspect ratios to accurately control biological samples. The photoresist must support the formation of intricate patterns with sharp sidewalls. SU-8 is widely used in BioMEMS due to its ability to create thick structures with vertical sidewalls [20]. Second, BioMEMS devices often operate in chemically reactive environments, such as bodily fluids (e.g., blood, urine) or laboratory reagents (e.g., hydrochloric acid (HCl), nitric acid (HNO_3)). The selected photoresist must resist degradation while maintaining its structural integrity [21]. Additionally, BioMEMS are frequently used in direct contact with biological cells, tissues, or fluids. Any toxic material or leachable chemicals from the photoresist could interfere with biological functions, which is why biocompatibility is essential [22]. Finally, many BioMEMS applications require long-term operation, where the materials must withstand mechanical stresses and environmental challenges. This necessitates photoresists with excellent mechanical properties, including rigidity and durability [20].

Given these stringent requirements, several photoresists are used in BioMEMS fabrication, each selected based on application-specific needs such as resolution, biocompatibility, and chemical stability. Several photoresists are used in BioMEMS fabrication, each selected based on application requirements such as resolution, biocompatibility, and chemical stability. SU-8 Photoresist: A negative-tone, epoxy-based photoresist that allows the fabrication of thick, high-aspect-ratio structures. It offers high mechanical strength, excellent chemical resistance, and biocompatibility. As previously discussed, it is commonly used in microfluidic devices, biosensors, and lab-on-a-chip systems [20, 21].

Dry-Film Photoresists: Suitable for applications requiring flat and uniform layers. They can be laminated to achieve different thicknesses, offering a streamlined fabrication process. These are often used in the fabrication of microfluidic moulds for BioMEMS [23]. AZ (e.g., AZ 4620) (Fig. 6.4) and SPR Series (e.g., SPR 220–7)

(Fig. 6.5) Photoresists: Positive-tone photoresists often used for developing precise microfeatures. These are typically employed in patterning sacrificial layers for BioMEMS structures and biomedical sensors [24, 25].

6.3 Optomechanical Microsystems: Microlenses and Optical Waveguides

Optomechanical microsystems are essential in modern photonics, enabling precise light manipulation for applications such as imaging, sensing, and optical communication. Two critical components in these systems are microlenses and optical waveguides, which play a pivotal role in directing and focusing light. Microlenses, often arranged in arrays, are used for collimation, focusing, and beam shaping, while optical waveguides guide light through confined structures, enabling efficient light transmission in photonic circuits and devices.

The fabrication of these components relies on advanced microfabrication techniques, particularly photolithography, thermal reflow, and soft lithography, which allow for precise control over geometry and optical properties. These methods have enabled the miniaturization and integration of microlenses and waveguides into microsystems, enhancing their performance and expanding their applications in fields such as MEMS, bioMEMS, and optical interconnects.

Microlens Fabrication Techniques

Thermal Reflow of Photoresist is one of the most widely used methods due to its simplicity and cost-effectiveness. In this process, photoresist cylinders are heated above their glass transition temperature, allowing surface tension to form smooth, spherical surfaces. While this method is efficient, it often struggles with controlling the geometry and uniformity of the microlenses. For instance, Lian et al. in 2014 demonstrated the fabrication of semi-ellipsoidal microlenses using two different photoresists with varying melting points, achieving high precision and reproducibility [26].

The thermal reflow process is used in microlens fabrication by coating a silicon substrate with photoresist (Fig. 6.6a), which is then patterned and developed to form cylindrical islands (Fig. 6.6b and Fig. 6.6c). The substrate is heated above the photoresist's glass transition temperature, allowing the material to flow and reshape under surface tension, forming spherical microlenses (Fig. 6.6d). Controlling the photoresist thickness and heating temperature is critical for achieving the desired microlens array geometry (Fig. 6.7).

In summary, Microlenses are fabricated using photoresists, which undergo chemical changes when exposed to light. One widely used method is thermal reflow, where cylindrical photoresist patterns are heated above their glass transition temperature, allowing surface tension to form spherical microlenses. This technique is particularly effective for creating high-fill-factor microlens arrays with precise curvature

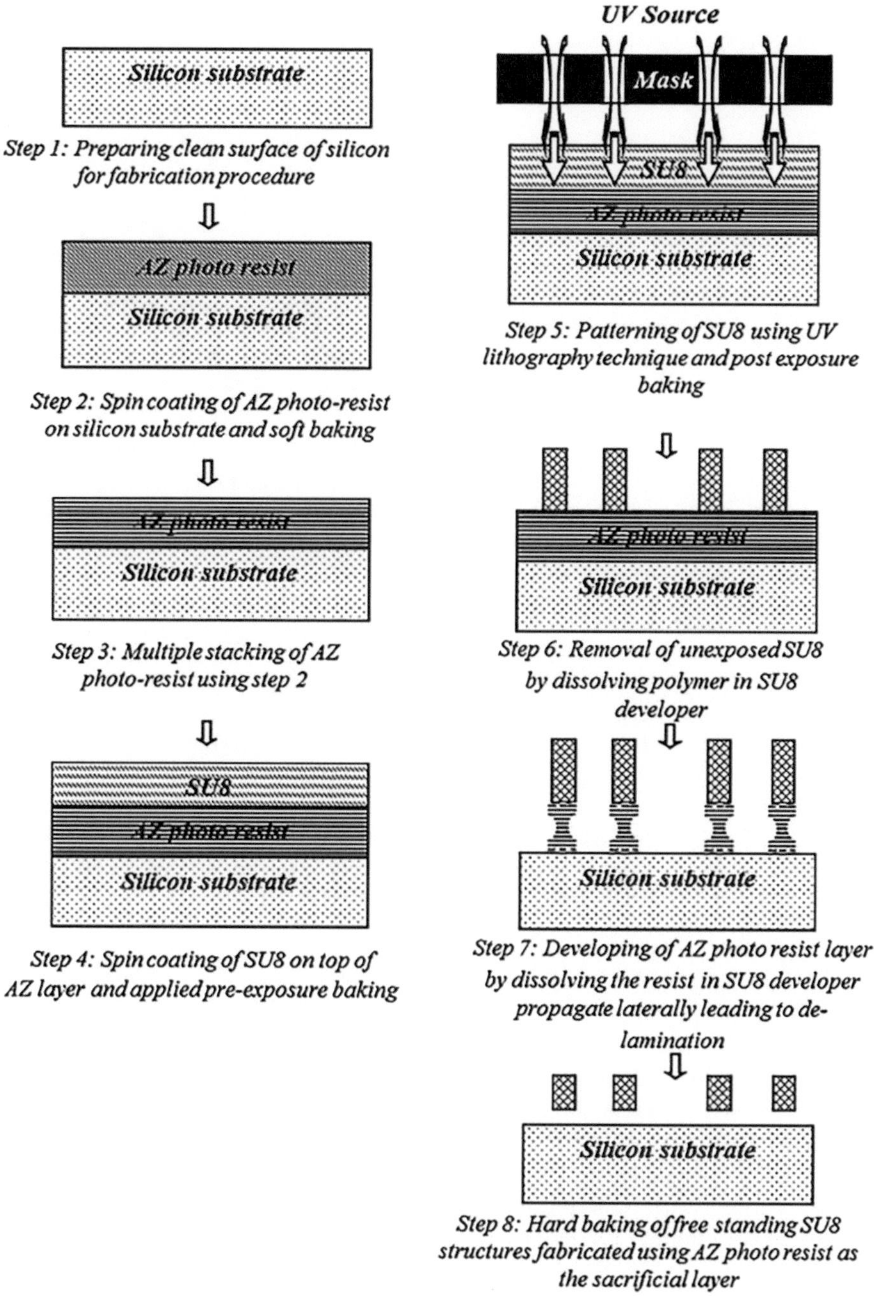

Fig. 6.4 Process flow for SU-8 structure fabrication and release using the AZ photoresist method [24]

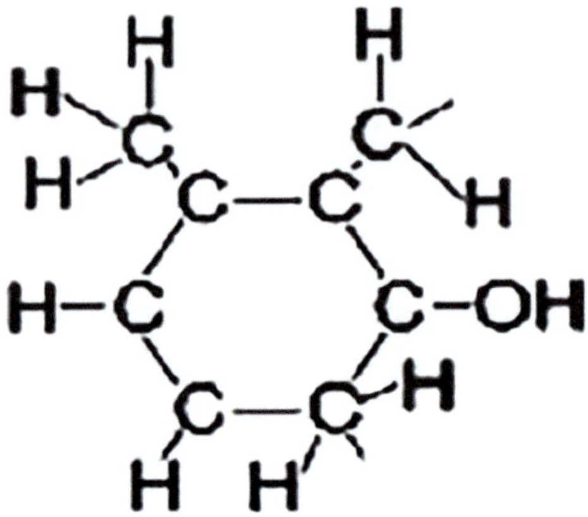

Fig. 6.5 Molecular structure of SPR 220–7 [25]

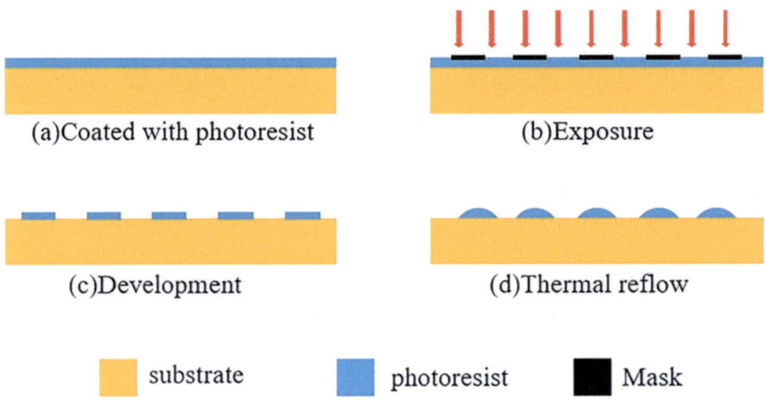

Fig. 6.6 Schematic of thermal reflow technology for fabrication of photoresist

control. For instance, studies have demonstrated the fabrication of semi ellipsoid microlenses using dual-photoresist systems with different melting points, enabling tilted microlens structures [28, 29].

Another approach is grayscale lithography, which uses a maskless exposure system to modulate light intensity and create 3D microlens structures directly on photoresist layers. This method offers high precision and flexibility in pattern design, as demonstrated in the fabrication of symmetric, asymmetric, and step-pattern microlens arrays [30]. Nanoimprinting with PDMS moulds is another cost-effective and scalable technique for large area microlens arrays. PDMS moulds replicate microlens patterns onto photoresist layers using UV curing or thermal processes, achieving high replication accuracy [31].

Laser-assisted fabrication is also employed, where femtosecond lasers directly ablate or modify photoresist surfaces to create microlens patterns. This method is suitable for high-precision applications requiring unique geometries, such as microlens

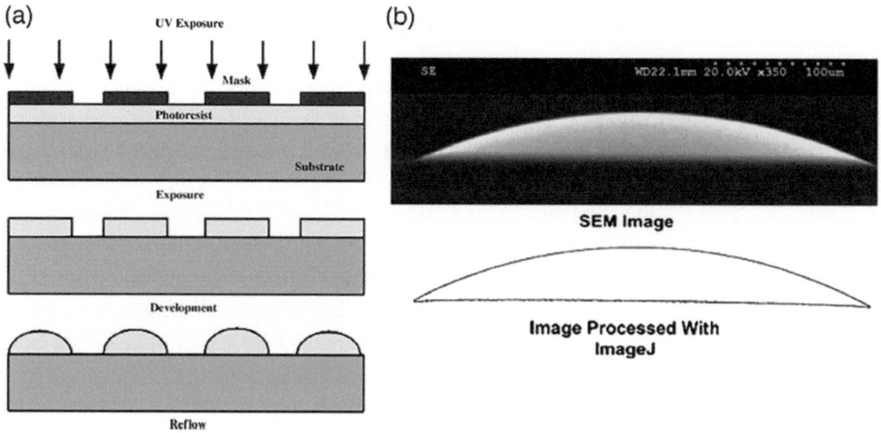

Fig. 6.7 **a** The fabrication process of MLAs by thermal reflows of photoresist: Firstly, the photoresist is spin-coated onto the substrate, and ultraviolet light passes through a circular mask to expose the photoresist. Then, heat treatment is performed. When heated to a temperature at which the photoresist enters a molten state, surface tension causes the photoresist to form a smooth spherical surface bounded by the mask pattern aperture, thereby creating the microlens surface structure. **b** SEM image of single microlenses. (Reproduced from Ashraf et al. [2008]) [27]

arrays for advanced optical systems [32]. Despite these advancements, challenges such as controlling geometric accuracy, uniformity, and processing time remain, prompting the exploration of hybrid methods that combine multiple techniques for improved cost-effectiveness and quality [33].

Optical Waveguide Fabrication Techniques

Optical waveguides are primarily fabricated using photolithography and etching processes, where photoresists like SU-8 and HSQ are widely utilized for their high resolution and etch resistance, making them ideal for integrated photonic circuits and lab-on-a-chip devices [34, 35]. Soft lithography offers an alternative approach, using PDMS moulds and UV-curable polymers to replicate tapered waveguides with high accuracy [36]. Dry film photoresists, such as SUEX, provide uniform thickness and are compatible with high-aspect-ratio structures, enabling the creation of precision waveguides like ridge gap designs for millimetre-wave applications with low insertion loss [37]. Additionally, advanced multilayer SU-8 processing allows for the fabrication of complex, high-aspect-ratio waveguide geometries, achieving exceptional optical performance, such as WR-3 band waveguides with minimal insertion loss (0.03 dB/mm) [6]. These techniques demonstrate the versatility and precision of photoresist-based methods in modern waveguide fabrication.

The fabrication process of the waveguide and bandpass filter involves a two-layer SU8 photolithographic process to overcome issues of air gaps formed during bonding of separate layers. Initially, SU8 50 photoresist is spun onto a 4-inch Si wafer, followed by prebaking at 65 °C and 95 °C to enable self-planarization and solvent drying. The wafer is then exposed to UV light using specific optical filters to achieve precise crosslinking of the SU8 layer. A second SU8 layer is subsequently added, aligned using alignment marks, and exposed under similar conditions, ensuring both layers are crosslinked and bonded during post-exposure baking. The final SU8 structure is developed and released from the Si substrate using EC solvent and sodium hydroxide. Metallization is achieved by oxygen plasma cleaning, a thin Cr adhesion layer, and thermal evaporation of a 2 μm thick silver layer, ensuring proper coating of sidewalls. The result is a fully joined, silver-coated SU8 structure forming one half of the waveguide or filter device.

To fabricate a tapered waveguide, a 30 μm thick photoresist layer is prepared through multiple spin-coatings of AZ9260 photoresist. The pattern, featuring a wider input and a narrower output terminal, is created using a mask aligner (Karl Suss MJB4), as illustrated in Fig. 6.8a. Oxygen/nitrogen plasma ashing is then performed on the photoresist, producing a vertically thicker section in the wider part and a thinner section in the narrower part, as shown in Fig. 6.8b. This patterned photoresist serves as the master mould for soft lithography. A polydimethylsiloxane (PDMS) mould is fabricated by curing the PDMS at 80 °C for 30 min (Fig. 6.8c). The final step of soft lithography involves imprinting the PDMS mould onto a UV-curable optical polymer to create a replica, as illustrated in Fig. 6.8d. The resulting tapered waveguide structure is represented schematically in Fig. 6.8e.

Materials and Processes

The choice of materials and processes plays a crucial role in microlens and waveguide fabrication, as they directly influence optical performance, durability, and fabrication efficiency. Commonly used materials include SU-8 photoresist, known for its high aspect ratio, light sensitivity, and thermal stability; PDMS, valued for its flexibility, thermal stability, and high light transmittance (>93%); PMMA, offering transparency, mechanical strength, and cost-effectiveness; and silica, which provides exceptional transparency (150 nm-4 μm), a high refractive index, and durability for high-performance applications.

Fabrication techniques such as spin coating, thermal reflow, soft lithography, and hot embossing are used to create precise microlenses, while characterization methods like optical microscopy and SEM ensure accuracy and quality. Each material is tailored to specific applications: PDMS and PMMA are suited for cost-effective, flexible designs, while silica excels in applications requiring superior optical and thermal properties. The final material choice depends on design requirements, processing methods, and the intended application.

Challenges and Innovations

Despite significant advancements, challenges remain in the fabrication of microlenses and waveguides. Surface roughness can affect optical performance, requiring precise control during fabrication. Alignment accuracy is critical for integrating microlenses and waveguides into microsystems, and material compatibility must be ensured to meet application requirements [33, 34].

Innovations in fabrication techniques have addressed some of these challenges. For instance, plasma ashing techniques have enabled the fabrication of tapered

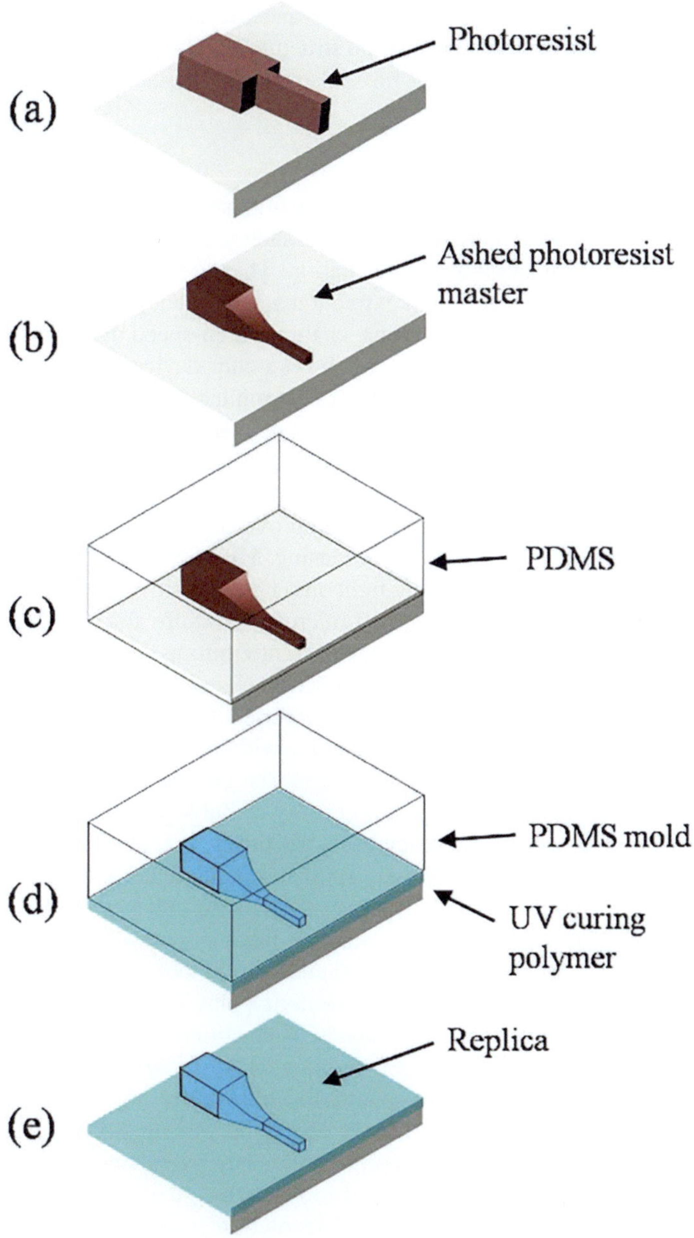

Fig. 6.8 Soft lithography process step: **a** patterning of photoresist by the conventional photolithography, **b** master preparation by plasma ashing, **c** PDMS mold fabrication, (d) imprinting with PDMS mold, and **e** final replica structure [35]

waveguides with precise dimensions, while advances in SU-8 and SUEX resists have facilitated the creation of high-aspect-ratio structures. Hybrid methods that combine thermal reflow with nanoimprinting have also improved scalability and precision [34, 35].

Applications

Microlenses and waveguides have diverse applications in optomechanical microsystems. Microlenses are used in imaging systems, such as cameras and microscopes, to enhance light collection and improve resolution. They are also employed in optical sensors and displays to improve light extraction and efficiency [35]. Optical waveguides are integral to telecommunications, enabling high-speed data transmission in photonic circuits. They are also used in lab-on-a-chip devices for biosensing and diagnostics, as well as in integrated photonics for miniaturized optical systems [36].

Microlenses and microlens arrays (MLAs) are essential micro-optical elements widely used in various fields due to their miniaturization, high integrability, and optical performance. In imaging, MLAs are applied in systems like microscopes, cameras, and biomedical imaging, enabling improved resolution, large field-of-view imaging, and 3D image reconstruction. In sensing, MLAs enhance the efficiency and sensitivity of light sensors by focusing light onto photosensitive areas, finding applications in astronomy, bio photonics, and microfluidic chips for cell detection and sorting. MLAs also improve the efficiency of light-emitting devices like OLEDs by enhancing light extraction and coupling efficiency, significantly boosting brightness and quantum efficiency. Additionally, MLAs are used in solar cells to reduce optical losses and improve power conversion efficiency without affecting device characteristics. These versatile applications demonstrate the critical role of microlenses in advancing imaging, sensing, lighting, and energy technologies [27].

Microlens arrays (MLAs) are crucial for structuring optical components, particularly for optical focusing, due to their ability to manipulate light efficiently and precisely. They are widely used in imaging systems to enhance resolution, depth of field, and light collection efficiency, making them essential for devices like 3D imaging systems, smartphone cameras, and endoscopic tools. In sensing and detection, MLAs improve light collection and sensitivity, playing a vital role in microfluidic chips for biological and chemical analysis. Their ability to collimate light ensures uniform illumination in projectors, LED lighting, and display backlights, while also enabling efficient optical coupling in systems like fibre optics and integrated photonic circuits by reducing coupling losses. In solar energy harvesting, MLAs focus sunlight onto solar cells, boosting power conversion efficiency by enhancing light trapping and reducing reflection. Biomedical applications include their use in optical coherence tomography (OCT) systems for high-resolution imaging and in laser-based surgical tools for precision. Emerging technologies such as augmented reality (AR), virtual reality (VR), and aerospace systems also leverage MLAs for immersive visual experiences and optical targeting. Liquid microlenses, a dynamic subset of MLAs, offer adjustable curvature for applications requiring variable focal lengths, such as adaptive optics in telescopes and tuneable lenses in cameras. Future advancements in hybrid fabrication techniques and novel materials will further enhance their potential, addressing challenges in uniformity, scalability, and cost while expanding their role in optical focusing and beyond.

6.4 Microfabricated Energy Devices: Batteries and Fuel Cells

Advances in photoresist-based patterning for energy devices: As technology advances, the demand for smaller, more efficient energy storage solutions has grown significantly. Microfabricated energy devices, such as microbatteries and microfabricated fuel cells (MFCs), play a crucial role in powering microelectromechanical systems (MEMS), portable electronics, and biomedical implants. These systems require on-chip energy storage to minimize power losses, improve efficiency, and reduce the need for external power sources [37].

One of the major challenges in developing these miniaturized energy devices is achieving precise structuring of their components, such as electrodes, electrolytes, and flow channels. To address this, photoresist-based lithography has become a fundamental tool. Photoresists, widely used in semiconductor fabrication, allow for high-resolution patterning, enabling the production of intricate microstructures with excellent control over shape, size, and placement. Beyond semiconductor applications, photoresist-based techniques are now widely used in the design and fabrication of energy storage devices, particularly in microbattery production and fuel cell development. This section explores how photoresist technology has contributed to advancements in these areas, highlighting recent progress, ongoing challenges, and future research directions. It also delves into recent advancements in photoresist-based microbattery and fuel cell fabrication, focusing on performance improvements, ongoing challenges, and future research directions [6, 38].

6.4.1 Photoresist in Microbattery Fabrication

One promising approach for integrating batteries into microfabricated systems is the use of photo-patternable radical polymer electrodes. These materials combine the electrochemical properties of organic radicals with the structuring capabilities of photoresists, enabling precise and scalable electrode fabrication.

A study by Cao et al. (2019) demonstrated how a modified SU-8 photoresist containing poly(2,2,6,6-tetramethyl-4-piperidinyl-N-oxyl methacrylate) (PTMA) and lithium perchlorate ($LiClO_4$) can be used to create microbattery electrodes with 10 μm resolution. The fabricated electrodes exhibited a charge/discharge potential of 0.68 V vs. Ag/AgCl and nearly 100% coulombic efficiency. However, despite these advantages, several challenges remain. The low areal capacity limits energy storage capabilities, while high interfacial impedance leads to energy losses during charge/discharge cycles. Furthermore, polymer swelling results in a 35% capacity loss after 60 cycles, reducing battery longevity.

Although these issues require further investigation, this study highlights the potential of radical polymer electrodes for on-chip energy storage. The ability to pattern such electrodes with high precision offers a promising pathway toward scalable and efficient microbatteries. Researchers are actively working on improving the electrical conductivity and mechanical stability of these electrodes to address existing limitations [37].

Another significant advancement in microbattery fabrication is the use of photoresist-assisted structuring to enhance energy density and performance. Lai et al. (2010) developed ultrahigh-energy–density lithium microbatteries using photolithography to create micropatterned electrodes with controlled porosity. This technique allows for optimized lithium-ion transport, ensuring fast and efficient charge transfer. The use of high-density sintered cathodes maximizes active material utilization, while the reduction of inactive material volume improves overall battery efficiency. These advancements resulted in energy densities of up to 675 Wh/L, making these batteries highly suitable for MEMS, medical implants, and wireless sensors.

In addition to energy storage performance, photoresist-based micropackaging techniques have significantly improved battery durability and mechanical stability. This makes these energy storage solutions more viable for long-term applications, where reliability and miniaturization are essential. Research in this field is focused on further optimizing electrode architecture and developing new materials to enhance the efficiency of microbatteries [6] (Fig. 6.9).

While microbatteries provide high energy storage density, their performance is limited by charge/discharge rates and cycle life. In contrast, microfabricated fuel cells (MFCs) offer continuous power generation, making them particularly useful for applications requiring sustained energy output over extended periods. The next section explores how photoresist-based lithography has contributed to MFC and PEM fuel cell fabrication.

6.4.2 Photoresist in Microfabricated Fuel Cells

Microbial fuel cells (MFCs) have emerged as a potential low-cost power source for disposable biosensors and diagnostic devices. One innovative approach involves using photoresist-based lithography to structure MFCs on paper substrates, ensuring precise fluid flow control and efficient energy generation.

Fraiwan et al. (2013) demonstrated how photoresist technology can be used to create hydrophobic barriers, defining anode and cathode compartments in paper-based MFCs [39]. Using UV lithography, they patterned hydrophilic and hydrophobic regions to ensure controlled liquid flow, prevent unintended reactant mixing, and

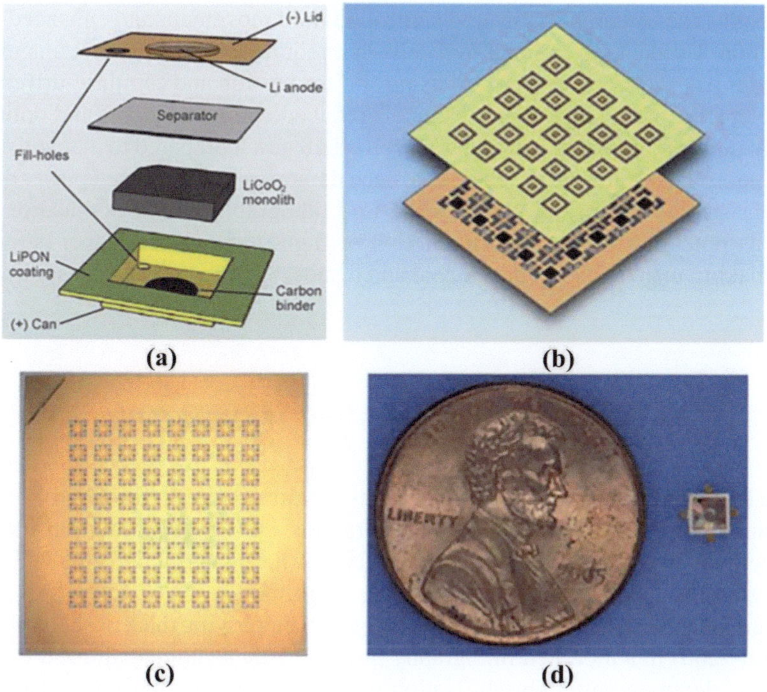

Fig. 6.9 **a** An exploded schematic view of the microbattery components. The cathode is attached inside a gold "can" using conductive carbon, while the lithium foil anode is attached to the copper lid. A film of insulating Li-P-O-N glass is deposited onto the can f langes. An intervening microporous separator is placed between the flange and the lid. The two halves of the cell are sealed using a fast-setting, photocured adhesive. **b** Schematic view of parallel assembly of multiple batteries enabled by the fabrication process. **c** Photograph of electroformed gold 64 can array. **d** Packaged microbattery compared to US penny [6]

facilitate effective proton exchange. This method also enables scalable and cost-effective fabrication, making it a practical solution for disposable energy applications. The resulting MFCs were capable of generating 5.5 $\mu W/cm^2$, with stacked configurations increasing voltage output by 1.9 $\times$. While their performance still falls short of commercial fuel cells, this research highlights how photoresist-based microfabrication can lead to practical, disposable power sources for biomedical applications.

The next step in this field involves improving power output by optimizing material selection and refining the design of microfluidic pathways [39].

Another emerging application of photoresist-based lithography is in vanadium co-laminar flow fuel cells, which use microfluidic channels to transport reactants. Jung and Ahn (2020) developed a microfabrication technique for creating precise and reproducible flow channels using SU-8 3035, a negative photoresist. By selectively converting hydrophilic paper regions into hydrophobic barriers through UV lithography, the researchers ensured controlled fluid movement and enhanced capillary action. Unlike conventional fuel cells that require external pumps, this method allows for passive reactant transport using surface tension and capillary-driven flow.

The precise patterning of flow channels reduced crossover losses and optimized reactant delivery, improving overall fuel cell efficiency. Compared to manually fabricated fuel cells, this microfabricated approach demonstrated improved power density and consistency. The use of photoresist-based lithography in this application highlights its potential for scalable production of high-performance, paper-based fuel cells, making them suitable for portable and disposable microelectronic applications [40].

6.4.3 Proton Exchange Membrane (PEM) Fuel Cells

Proton Exchange Membrane (PEM) fuel cells require precisely structured electrodes to maintain high efficiency and minimal energy losses. However, traditional fabrication methods often struggle with adhesion issues and material deformation, which can negatively impact fuel cell performance over time.

Shah et al. (2003) addressed these challenges by employing thick, viscous photoresist (AZ-9260) in a modified lithography process. This technique allowed for high-precision electrode patterning on Nafion membranes, improving ion transport and structural stability. The use of PDMS shadow mask fabrication ensured uniform metal deposition, preventing material spreading and enhancing performance. Additionally, gas permeability was improved, reducing ohmic losses and increasing the overall efficiency of the fuel cell. These advancements led to fuel cells achieving an open-circuit voltage of 840 mV and a power density of 0.375 mW/cm^2. This research demonstrates how photoresist-based lithography is enabling the development of scalable, high-performance PEM fuel cells for portable power applications. Further research aims to enhance durability and integrate these systems with other microfabricated energy technologies to improve overall performance [41].

Table 6.3 provides a clear comparison of microfabricated energy storage solutions, showcasing key advancements in efficiency, charge capacity, and fabrication techniques. While photoresist-based microfabrication has significantly improved the performance and scalability of microbatteries and fuel cells, there are still important challenges to overcome. Researchers continue to focus on enhancing charge transport, extending fuel cell lifespan, and increasing battery capacity to make these technologies more reliable and efficient. The next section explores future directions and emerging trends that could further advance photoresist-based energy device fabrication.

Despite the significant progress made in photoresist-based energy device fabrication, several challenges remain, including but not limited to enhancing charge transport, improving fuel cell longevity, and increasing battery capacity. As research progresses, advancements in photoresist chemistry and microfabrication techniques

Table 6.3 Summar of performance Metrics for Photoresist based Micro Fabricated energy devices [37, 41]

Method	Key characteristics
Modified SU-8 photoresist containing poly(2,2,6,6-tetramethyl-4-piperidinyl-N-oxyl methacrylate) (PTMA) and lithium perchlorate ($LiClO_4$)	- Charge/discharge potential: 0.68 V versus Ag/AgCl - Coulombic efficiency: Nearly 100% - Areal capacity: 1×10^{-5} mAh·cm^{-2} (only 1/8 of theoretical capacity) - Capacity retention: 65% after 60 cycles
Photoresist-assisted electrode structuring for high-energy–density lithium microbatteries	- Energy density: 675 Wh/L - Charge transfer efficiency: 99.5% - Microbattery thickness: < 1 mm (ideal for MEMS integration)
UV lithography for patterning hydrophilic and hydrophobic regions in paper-based microbial fuel cells (MFCs)	- Power output: 5.5 μW/cm^2 - Voltage boost with stacking: 1.9 × increase in series connection - Operational longevity: Stable performance across multiple cycles
Photoresist-assisted microfabrication in vanadium co-laminar flow fuel cells	- Improved reactant transport efficiency - Reduced crossover losses - Higher fuel utilization compared to traditional designs
AZ-9260 photoresist for high-precision Nafion membrane electrode patterning in PEM fuel cells	- Open-circuit voltage: 840 mV - Power density: 0.375 mW/cm^2 - Enhanced gas permeability for improved efficiency

will continue to push the boundaries of miniaturized energy storage, leading to more efficient, scalable, and durable energy devices [6, 37, 41].

6.5 Role of HSQ and OSTE Polymers in Specialized Microsystems Applications

Case Study 1: Hydrogen Silsesquioxane (HSQ) in High-Resolution Optical MEMS

HSQ's unique properties as a high-resolution, inorganic negative-tone resist have enabled breakthroughs in optical MEMS, particularly for sub-100 nm photonic crystals and waveguide gratings. In one application, HSQ was used to pattern nanoscale Bragg gratings on silicon-on-insulator (SOI) substrates for integrated optical gyroscopes [2, 42]. The process involved electron-beam lithography to define 80 nm-wide grating lines with a pitch of 220 nm, achieving a refractive index contrast of 2.5 through selective oxidation. These gratings demonstrated a polarization-dependent loss of < 0.1 dB/cm, critical for reducing noise in Coriolis force detection systems.

Post-patterning, HSQ's thermal stability (up to 900 °C) allowed direct integration with CMOS-compatible annealing processes, eliminating delamination risks observed with organic resists like SU-8 [42].

Case Study 2: OSTE Polymers in Microfluidic Organ-On-Chip Systems

OSTE polymers have revolutionized bioMEMS by enabling leakage-free, multimaterial integration of nanoporous membranes into organ-on-chip platforms. In a 2024 study, OSTE(+), a dual-cure thiol-ene-epoxy variant, was employed to bond 200 nm-pore alumina membranes to microfluidic channels for a lung-on-chip model [2, 42]. The first UV cure created a compliant OSTE(+) layer (Young's modulus: 0.5 MPa), which conformed to the membrane's nanoporous surface at room temperature. The second thermal cure (80 °C, 1 h) then crosslinked epoxy groups, achieving a bond strength of 2.1 MPa without compromising membrane integrity. This approach facilitated simultaneous air–liquid interface culture and real-time cytokine sampling, with < 3% crosstalk between epithelial and endothelial channels [42].

OSTE's built-in thiol groups further allowed direct grafting of collagen-mimetic peptides via thiol-ene click chemistry, enhancing cell adhesion by 60% compared to plasma-treated PDMS [9].

Case Study 3: Off-Stoichiometry Thiol-Ene (OSTE) Polymers with Allylsilanes for Microfluidic and Silicon Integration

A novel class of off-stoichiometry thiol-ene (OSTE) polymers incorporating silane groups via allylsilanes (OSTE-AS) has been developed to improve bonding with silicon wafers without surface pretreatment. By modifying thiol monomers with allylsilanes, the OSTE-AS polymers achieve superior adhesion, optimized mechanical properties, and enhanced chemical resistance. The polymer formulation was fine-tuned to balance tensile strength (50 MPa), hardness (Shore D: 83), Young's modulus (1.76 GPa), and thermal stability (glass transition at 92 °C, degradation at 320 °C). The dielectric constant (4.2) and optical transparency make OSTE-AS suitable for microelectronic applications, while its strong adhesion to silicon enables direct spin-coating integration with standard fabrication techniques.

Compared to previous OSTE modifications, this approach provides a flexible strategy for producing thiol- or allyl-excess polymers, facilitating further functionalization. Demonstrating its practicality, a microfluidic system was successfully fabricated using OSTE-AS bonded to silicon, validating its potential in lab-on-chip, biosensor packaging, and microfluidic integration for biomedical and analytical applications [26] (Fig. 6.10).

Case Study 4: Off-Stoichiometry Thiol-Ene (OSTE) Polymers for Microfluidic Device Prototyping

A novel polymer platform based on off-stoichiometry thiol-enes (OSTEs) has been developed to bridge the gap between research prototyping and commercial microfluidic device production. Utilizing UV-curable thiol-ene chemistry with off-stoichiometry ratios, OSTE polymers enable one-step surface modifications, tuneable

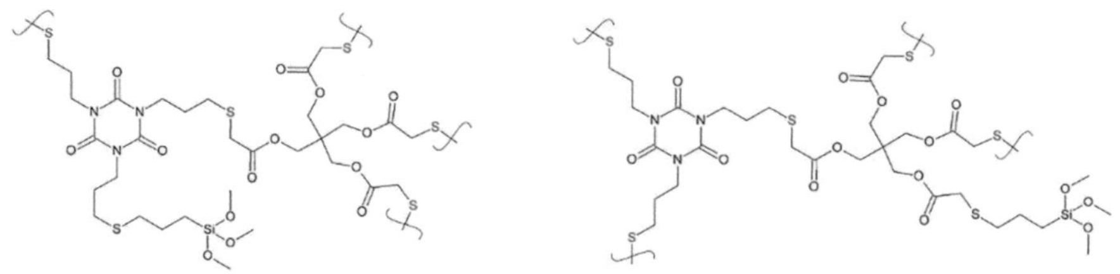

Fig. 6.10 The silane-containing fragments of OSTE-MS and OSTE-AS polymers. The sulfur bridges between monomer fragments are indicated in red. The sulfur bridges between monomer fragments and silanes are indicated in blue [8]

mechanical properties, and leak-free sealing through direct UV-bonding. The material retains the rapid prototyping and ease of processing of PDMS while offering mechanical and chemical properties similar to commercial-grade thermoplastics.

OSTEs can be cast using SU-8 moulds with standard UV lamps, allowing precise surface modifications via stencil masks and single-step UV bonding. Demonstrating its versatility, OSTE was used to fabricate microfluidic chips featuring patterned hydrophobic stops, pneumatic valves through UV-laminated stiff and elastomeric layers, and chip-to-world connectors. This innovation presents a scalable alternative to PDMS, enhancing manufacturability and performance for lab-on-chip applications [43].

Case Study 5: High-Temperature Processed Silicon Nanocrystals for Scalable Applications

Silicon nanocrystals (Si-NCs) exhibit size-dependent optical and electronic properties, making them valuable for applications in sensors, photovoltaics, and biological systems due to their inherent non-toxicity. However, achieving high-purity, monodisperse Si-NCs has been a persistent challenge. A scalable synthesis method utilizing high-temperature processing of hydrogen silsesquioxane (HSQ) provides a reliable route to producing pure, near-monodisperse nanocrystals. These particles are liberated through HF etching and can be functionalized via multiple techniques to tailor their surface chemistry for specific applications.

This study details the synthesis protocol, functionalization approaches, and key characterization techniques, establishing a reproducible framework for integrating Si-NCs into advanced optoelectronic and biomedical technologies [44].

Conclusions

The evolution of photoresist materials has been pivotal in advancing Micro-Electro-Mechanical Systems (MEMS), enabling breakthroughs across sensors, actuators, and specialized microsystems. In MEMS devices, photoresists such as SU-8, SPR 220–7, and Ordyl P-50100 have facilitated the creation of high-aspect-ratio structures, precise metallization patterns, and robust 3D packaging solutions. These innovations enhance the performance of accelerometers and gyroscopes, improving sensitivity, noise reduction, and thermal stability. Emerging materials like molecular and hybrid resists promise finer feature sizes and adaptive functionalities, though challenges such as stiction and residual stress persist. Future advancements aim to leverage smart photoresists for applications in biomedical, automotive, and consumer electronics sectors.

Transitioning to BioMEMS and microfluidics, SU-8's chemical stability, heat resistance, and biocompatibility make it ideal for lab-on-a-chip devices, which miniaturize diagnostic and research processes. Complemented by OSTE polymers and dry-film photoresists, these systems offer flexibility in multi-material integration, prioritizing durability and biosafety for biological applications. These technologies are revolutionizing point-of-care diagnostics, drug delivery, and tissue engineering, with ongoing refinements in materials and fabrication techniques.

In optomechanical microsystems, photoresists like SU-8 and PDMS, combined with thermal reflow, grayscale lithography, and femtosecond laser writing, underpin the fabrication of microlenses and optical waveguides. These components are critical for imaging, sensing, and telecommunications, though challenges in geometric accuracy and surface roughness remain. Innovations such as liquid microlenses and hybrid methods aim to enhance scalability, positioning these systems for future roles in artificial intelligence, medical devices, and next-generation photonics.

Microfabricated energy devices, including microbatteries and fuel cells, benefit from photoresist-assisted lithography for high-precision electrode structuring and microfluidic channel integration. Studies by Jiang and Weng (2019) demonstrate improved conductivity and capacity through 2D materials, while Wainright et al. (2003) and Jung and Ahn (2020) highlight advancements in fuel cell miniaturization. Overcoming challenges in charge transport, material stability, and electrode degradation will be crucial for scaling these technologies for compact electronics and biomedical implants [45, 46].

Specialized photoresists like HSQ and OSTE polymers further exemplify MEMS innovation. HSQ enables nanoscale Bragg gratings for optical MEMS and scalable synthesis of pure silicon nanocrystals, while OSTE polymers revolutionize bioMEMS through organ-on-chip systems and simplified silicon bonding without pretreatment. Their tuneable properties bridge prototyping and commercial production, underscoring their versatility in microfluidic and biomedical applications.

Collectively, photoresist technology continues to drive MEMS advancements, addressing fabrication challenges and enabling multi-functional integration across diverse fields. By refining material chemistry and lithographic techniques, future innovations will expand the role of MEMS in next-generation optical, energy, and biomedical technologies.

References

1. M. Hassaan, U. Saleem, A. Singh, A.J. Haque, K. Wang, Materials 17, (2024)
2. N. Wangler, L. Gutzweiler, K. Kalkandjiev, C. Müller, F. Mayenfels, H. Reinecke, R. Zengerle, N. Paust, J. Micromech. Microeng. 21, (2011)
3. A.T. Giannitsis, Est. J. Eng. 17, (2011)
4. J. Chung, H.Y. Hwang, Y. Chen, T.Y. Lee, "Sens. Actuators, B" 254, 542–550 (2018)
5. H. Xiong, Z. Wang, 2018 IEEE SENSORS 1–3 (2018)
6. W. Lai, C.K. Erdonmez, T.F. Marinis, C.K. Bjune, N.J. Dudney, F. Xu, R. Wartena, Y.M. Chiang, Adv. Mater. 22, (2010)
7. A. Fraiwan, S. Choi, Biosens. Bioelectron. **83**, 27–32 (2016)
8. K. Puchnin, D. Ryazantsev, E. Latipov, V. Grudtsov, A. Kuznetsov, Polymers 15, (2023)
9. D.C. Smallwood, P. McCloskey, C. O'Mathuna, D.P. Casey, J.F. Rohan, Microsyst. Nanoeng. 7, (2021)
10. W.C. Wilson, G.M. Atkinson, NTRS L-19339 (2007)
11. Y.Y. Chen, Y. Sun, Q.B. Zhu, B.W. Wang, X. Yan, S. Qiu, Q.W. Li, P.X. Hou, C. Liu, D.M. Sun, H.M. Cheng, Adv. Sci. 5, (2018)
12. M.R. Jamiruddin, B.A. Meghla, D.Z. Islam, T.A. Tisha, S.S. Khandker, M.U. Khondoker, M.A. Haq, N. Adnan, M. Haque, Life 12, (2022)
13. F. Zhu, A. Wigh, T. Friedrich, A. Devaux, S. Bony, D. Nugegoda, J. Kaslin, D. Wlodkowic, Environ. Sci. Technol. **49**, 14570–14578 (2015)
14. B. Gokce, I. Akcok, A. Cagir, D. Pesen-Okvur, Eur. J. Pharm. Sci. 155, (2020)
15. S. Haeberle, R. Zengerle, Lab Chip 7, (2007)
16. C.-H. Lin, G.-B. Lee, B.-W. Chang, G.-L. Chang, J. Micromech. Microeng. **12**, 590–597 (2002)
17. S. Pashayev, R. Lhermerout, Christophe Roblin, E. Alibert, J. Barbat, R. Desgarceaux, R. Jelinek, E. Chauveau, S.ı. Tahir, V. Jourdan, R. Jabbarov, F. Henn, A. Noury, Microfluid. Nanofluid. 28, 25 (2024)
18. R. Ruska, A. Sarakovskis, E. Zutis, G. Paidere, I. Vozny, J. Cipa, J. Gabrusenoks, T. Freimanis, R. Zalubovskis, A. Anspoks, Materials 17, (2024)
19. X. Liu, S. Li, SLAS Technol. **25**, 151–161 (2020)
20. V.N. Vernekar, D.K. Cullen, N. Fogleman, Y. Choi, A.J. García, M.G. Allen, G.J. Brewer, M.C. LaPlaca, J. Biomed. Mater. Res. Part A **89A**, 138–151 (2008)
21. G. Gharib, İ. Bütün, Z. Muganlı, G. Kozalak, İ. Namlı, S.S. Sarraf, V.E. Ahmadi, E. Toyran, A.J. van Wijnen, A. Koşar, Biosensors 12, (2022)
22. W.-m. Zhou, Y.-y. Yan, Q.-r. Guo, H. Ji, H. Wang, T.-t. Xu, B. Makabel, C. Pilarsky, G. He, X.-y. Yu, J.-y. Zhang, J. Nanobiotechnol. 19, (2021)
23. M.M. Roos, A. Winkler, M. Nilsen, S.B. Menzel, S. Strehle, International Journal of Precision Engineering and Manufacturing-Green Technology **9**, 43–57 (2021)
24. K.H. Lau, A. Giridhar, S. Harikrishnan, N. Satyanarayana, S.K. Sinha, Microsyst. Technol. **19**, 1863–1871 (2013)
25. E. Koukharenko, M. Kraft, G.J. Ensell, N. Hollinshead, J. Mater. Sci. Mater. Electron. **16**, 741–747 (2005)
26. Z.-J. Lian, S.-Y. Hung, M.-H. Shen, H. Yang, Microelectron. Eng. **115**, 46–50 (2014)
27. S. Cai, Y. Sun, H. Chu, W. Yang, H. Yu, L. Liu, Microsc. Res. Tech. **84**, 2784–2806 (2021)
28. R. Winston, P. Schreiber, R.J. Koshel, S. Kudaev, P. Dannberg, U.D. Zeitner, Homogeneous LED-illumination using microlens arrays, Nonimaging Optics and Efficient Illumination Systems II, 2005
29. H. Ottevaere, R. Cox, H.P. Herzig, T. Miyashita, K. Naessens, M. Taghizadeh, R. Völkel, H.J. Woo, H. Thienpont, J. Opt. A Pure Appl. Opt. **8**, S407–S429 (2006)
30. H. Yang, M.A.M. Gijs, Chem. Soc. Rev. **47**, 1391–1458 (2018)
31. Y.-J. Weng, Y.-C. Weng, J.-C. Huang, Y.-C. Wong, S.-Y. Yang, H.-K. Liu, Fabrication of Microlens Arrays by Using Nano-Particle Fluid Imprinting Technology. International Conference on Signal Processing Systems **2009**, 963–966 (2009)

32. S. Matsuo, T. Miyamoto, T. Tomita, 2005 Pacific Rim Conference on Lasers & Electro-Optics 1731–1732 (2005)
33. S.-Y. Hung, C.-H. Hung, 2012 12th International Conference on Numerical Simulation of Optoelectronic Devices 67–68 (2012)
34. X. Shang, M. Ke, Y. Wang, M.J. Lancaster, IEEE Trans. Terahertz Sci. Technol. **2**, 629–637 (2012)
35. H.-H. Kim, J.-H. Han, D.-H. Lee, B.-H. O, S.-G. Lee, E.-H. Lee, S.-G. Park, Microelectron. Eng. 88, 2721–2724 (2011)
36. M. Dubov, S.R. Natarajan, J.A.R. Williams, I. Bennion, In: Commercial and Biomedical Applications of Ultrafast Lasers VIII 6881, (2008)
37. L. Cao, G. Fang, H. Cao, X. Duan, Langmuir **35**, 16079–16086 (2019)
38. K. Jiang, Q. Weng, Chemsuschem **13**, 1420–1446 (2019)
39. A. Fraiwan, S. Mukherjee, S. Sundermier, H.-S. Lee, S. Choi, Biosens. Bioelectron. **49**, 410–414 (2013)
40. D.-G. Jung, Y. Ahn, J. Power Sources 451, (2020)
41. K. Shah, W.C. Shin, R.S. Besser, J. Power. Sources **123**, 172–181 (2003)
42. F. Saharil, F. Forsberg, Y. Liu, P. Bettotti, N. Kumar, F. Niklaus, T. Haraldsson, W. van der Wijngaart, K.B. Gylfason, J. Micromech. Microeng. 23, (2013)
43. C.F. Carlborg, T. Haraldsson, K. Öberg, M. Malkoch, W. van der Wijngaart, Lab Chip 11, (2011)
44. R.J. Clark, M. Aghajamali, C.M. Gonzalez, L. Hadidi, M.A. Islam, M. Javadi, M.H. Mobarok, T.K. Purkait, C.J.T. Robidillo, R. Sinelnikov, A.N. Thiessen, J. Washington, H. Yu, J.G.C. Veinot, Chem. Mater. **29**, 80–89 (2016)
45. P. Zhang, F. Wang, M. Yu, X. Zhuang, X. Feng, Chem. Soc. Rev. **47**, 7426–7451 (2018)
46. J.S. Wainright, R.F. Savinell, C.C. Liu, M. Litt, Electrochim. Acta **48**, 2869–2877 (2003)

Chapter 7
High-Aspect-Ratio Lithography for Microsystems

This chapter focuses on high-aspect-ratio lithography, discussing the requirements for high-aspect-ratio patterning and the properties and applications of SU-8 and related resists. It highlights the role of HSQ in high-resolution, high-aspect-ratio structures and addresses the challenges in deep lithography, such as resist swelling, adhesion loss, and cracking. Applications in MEMS and other microsystems are also presented.

7.1 Requirements for High-Aspect-Ratio Patterning

High-aspect-ratio lithography (HARL) is a vital technology in modern microsystems manufacturing. It enables the creation of three-dimensional structures essential for microelectromechanical systems (MEMS), microfluidics, and optical devices [1]. The ability to fabricate HAR structures directly affects device performance and factors like sensitivity, efficiency, and integration density [1]. This chapter introduces the topic, including the basic requirements, key performance metrics, and challenges in this field. The most important key features are visualized in Fig. 7.1.

7.1.1 Key Features and Requirements

The aspect ratio (AR) of a microstructure is defined as the ratio between the structure's height (h) and critical dimension (CD), which can be expressed as AR = h/CD [2]. Higher aspect ratios enable more compact and efficient devices by increasing surface area utilization and improving functional properties. While conventional lithography processes typically operate at AR < 5:1, HARL targets AR of 10:1 and larger. State-of-the-art techniques can reach AR > 25:1 using resists like SU-8 and hydrogen

K. Wang, *Photoresist Technology in Microsystems: Principles, Processes and Applications*, Springer Series in Advanced Microelectronics 15,
https://doi.org/10.1007/978-981-95-1606-3_7

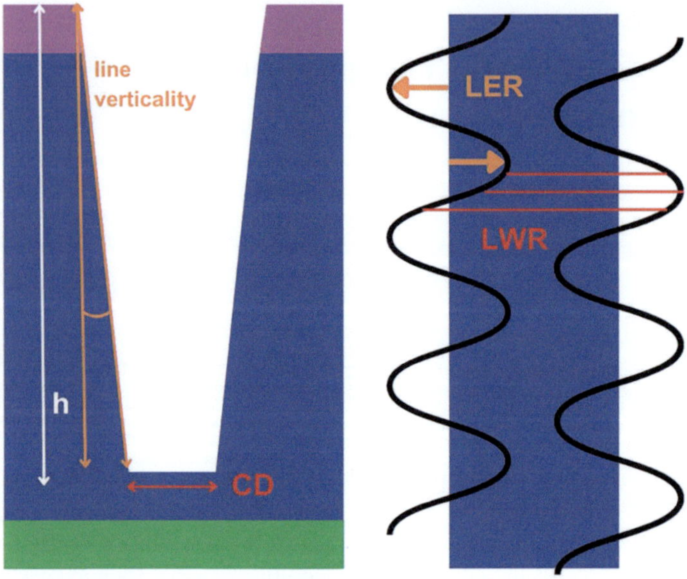

Fig. 7.1 Visualization of critical dimension (CD), line verticality, line edge roughness (LER), line width roughness (LWR)

silsesquioxane (HSQ), which provide the required mechanical stability and chemical resistance [3]. Achieving high-aspect ratios requires precise control over exposure, development, and etching processes to maintain structural fidelity.

The feature resolution describes the smallest dimension (CD) that can be reliably transferred during lithography. Achieving a high resolution is crucial for applications demanding miniaturization. Unlike planar lithography processes, where wavelength limitations primarily determine resolution, HARL faces additional constraints related to energy penetration, scattering effects, and development dynamics [4].

Radiation scattering within the bulk material in deep resist layers leads to dose variations that degrade the resolution. [5] Furthermore, diffusion-limited development kinetics affect the dissolution rate of exposed regions, potentially causing distortions. Mechanical deformation during processing, including stress relaxation and resistance swelling, further complicates precise control over feature resolution [6]. Therefore, the resolution in HARL systems mainly depends on radiation scattering within the resist bulk, developer transport dynamics, and mechanical deformation during processing. These three factors compete with one another.

Furthermore, the verticality and smoothness of microstructure sidewalls are essential for ensuring uniform material deposition during subsequent fabrication steps such as etching or metallization. Deviations from the verticality of the applied pattern, such as tapering or footing, usually arise from uneven exposure doses, resist composition gradients, or development rate disparities [4, 7].

The surface roughness, often called *line edge roughness* (LER), must also be kept small to minimize optical scattering losses and mechanical stress concentrations. Sidewall roughness can be produced by many factors, such as resist material properties, exposure wavelength, post-exposure treatment, and development conditions [8].

Photoresists in HARL must have high etch selectivity. In this context, etch selectivity describes the ratio of the substrate etch rate to the resist's erosion rate.[9] A high etch selectivity ensures that HAR patterns are precisely transferred to the underlying materials. Selectivity requirements escalate with increasing AR [9].

Resolution, AR, and throughput must be balanced to meet industrial scalability requirements while considering process fluctuations. HARL processes must tolerate exposure intensity and focus variations while maintaining critical dimensions, as HAR lithography machines often work at the limits of their depth of field. [10] Furthermore, post-processing steps such as development and drying must consider the increased mechanical fragility of HAR structures without sacrificing yield.

7.1.2 Fundamental Challenges

One of the fundamental challenges in HAR patterning is achieving uniform exposure energy deposition within thick photoresist layers. [4] High-energy radiation sources, such as X-rays or high-voltage electron beams, attenuate this effect, causing tradeoffs between penetration depth and resolution. [11] Nonuniform absorption can cause incomplete crosslinks or development lag.

Dealing with developer chemicals in deep and narrow trenches follows complex kinetic movements. Therefore, Developer Transport and Dissolution Dynamics is another challenge in HARL. Concentration gradients form along the resist depth, potentially causing inhomogeneous dissolution rates. [12] While this can lead to residual resists at the base in negative-tone resists, positive-tone resists may exhibit undercut profiles.

Polymeric resists absorb solvents during development, causing volumetric swelling that destroys high-aspect-ratio features. [6] Swelling-induced strain becomes crucial at aspect ratios larger than 20:1, where capillary forces during drying further support structural collapse. [4]

Adhesion between the photoresist layer and the substrate material must withstand shear stresses during several development and rinse cycles. Weak interfacial bonds can lead to delamination, especially at resist-substrate interfaces where stress concentrations are highest. [13]

Internal stresses from thermal expansion (CTE) mismatches, shrinkage, or resist swelling generate tensile forces that can fracture HAR structures. As these stresses usually scale with the thickness of the photoresist layer, crack propagation is a critical failure mode in narrow and tall features. [4]

Considering an optimal resist profile, transferring HAR patterns into a substrate still requires etching processes with near-ideal anisotropic characteristics. Isotropic etching techniques would undercut resist features, lowering the effective aspect ratio. At the same time, erosion in the sidewalls of the resist during prolonged etching can further degrade the pattern's fidelity. [6]

7.2 SU-8 and Related Resists: Properties and Applications

SU-8 is an epoxy-based negative photoresist that produces High-Aspect Ratio (HAR) microstructures in photolithography [3]. This characteristic makes it widely used in MEMS applications and is particularly popular for bio-MEMS due to its good biocompatibility [14]. The main attribute of SU-8 is the capability to create a 3D network of cross-links, given that the irradiation quantity is sufficiently elevated. The outcome is a resist mask that is mechanically and thermally stable [3].

The key performance metrics of patterning with SU-8 are ratios up to 20:1 or even higher in optimized conditions [15] and vertical sidewalls with angles close to 90° [16]. When using proper formulation, high resolution can also be achieved in the order of sub-50 nm patterns [17]. SU-8's chemical properties influence these metrics. The main chemical characteristic of the SU-8 is that it is a negative tone-resist, meaning it becomes insoluble when exposed to light. It is also chemically amplified, so a tiny amount of light exposure triggers a chain reaction, strengthening the material [4]. The key chemical components are the acid-labile groups and the photoacid generator. Its high Young's modulus (~4–5 GPa) reflects its good mechanical strength. Its thermal stability is demonstrated by a transition temperature of approximately 200 °C and a degradation temperature of approximately 380 °C. The resist is commercially available in various formulations, making it suitable for applications with film thicknesses ranging from 2 to 300 μm [12]. The lithographic process with SU-8 begins with selecting the appropriate variant to achieve the expected thickness. The process's spin velocity can range from 1000 to 3000 rpm depending on the variant's viscosity. Since SU-8 is hydrophobic, it has poor wetting behavior on hydrophilic substrates like oxidized silicon wafers, leading to adhesion challenges. Research suggests that Ti and TiO_2 substrates are appropriate for a stronger adhesion, while Ni should be avoided [12]. Soft-baking is typically performed at 95 °C to eliminate solvents and enhance material adhesion. Exposure is then performed using a UV lamp and mask or laser scanning. This is crucial, as the resist's UV absorption at wavelengths above 350 nm can affect thick SU-8 layers, causing nonvertical sidewalls and pronounced T-shaped profiles. After exposure, the post-bake step enhances cross-linking in the irradiated regions and reinforces structural stability. Finally, the wafer is submerged in a developing solution to dissolve unexposed SU-8, completing the development phase as seen on Fig. 7.2.

SU-8 can be used to obtain HAR patterns using different types of exposures, such as UV, X-ray, electron beam, and proton beam [4]. With UV lithography, ratios can be up to 20:1, and the thickness can range from 200 to 700 μm. The X-ray exposure, with short wavelengths, can penetrate thick layers and create vertical sidewalls, delivering results of ratios up to 100 [3]. However, the lack of accessibility to

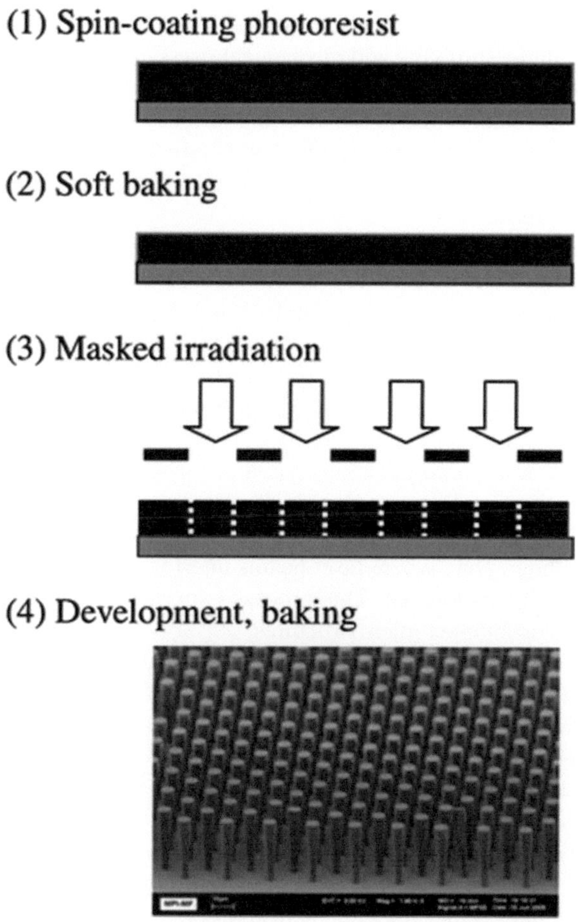

Fig. 7.2 Photolithographic steps with the negative photoresist SU-8

synchrotron radiation equipment restricts it from the broad public. Another possibility is the use of electron beams, even though SU-8 was initially developed for optical applications. The photoresist offers a higher sensitivity to e-beam, being suitable for nanopatterning. While SU-8 can be patterned in thicker layers using e-beam lithography, achieving high resolution is challenging beyond 200 nm due to limited penetration depth and scattering effects. Another limitation of this method is electron scattering, which causes unintended exposure of resist molecules beyond the targeted area, reducing overall resolution. Finally, proton beam lithography uses high-energy ions to penetrate the resist deeply and generate a well-defined path. The big difference between the ion and electron mass leads to negligible deviation in the ion's trajectory from its path. This feature allows the ion beam to create a smooth, vertical sidewall with high packing density, properly fabricating HAR structures.

Using SU-8 to achieve HAR has specific challenges. First, planarization defects, such as surface unevenness, often occur in resist layers around 1 mm thick, with height variations ranging from 10 to 100 μm [18]. Furthermore, during soft baking, volume shrinking and mechanical stress can occur due to solvent removal. The stress is proportional to the film thickness. Also, the long development times can damage fine features due to the swelling of cross-linked regions [12]. Ultimately, due to the decrease in lateral size of the structures, a collapse can occur when adhesive forces between features exceed the mechanical forces needed to maintain their upright position.

Some unconventional lithography techniques are used to overcome these challenges and create deep sub-micron or tall three-dimensional structures, achieving the high-aspect-ratio microstructures. For example, the inclined UV exposure creates unique microstructures at various angles [19]. Another technique is the back-side UV exposure; given that the photoresist layers are thicker, the UV exposure dose is not uniform from the top to the bottom, creating a narrower structure at the bottom and a wider one at the top. The effect is reversed by exposing from the backside through a UV-transparent substrate. Another non-traditional UV lithography that can be applied is drawing lithography, which utilizes the mechanical pulling of a soft-baked SU-8 layer to create a tall and thin microneedle [20]. Finally, the holographic interference lithographic, where laser beams interfere with each other to create periodic patterns on the photoresist. This process is mask-less and relies on interference patterns to form complex microstructures.

While SU-8 is used for high-aspect-ratio lithography, chemically amplified resists (CARs) are an alternative for applications requiring higher resolution and etch resistance. Initially developed for deep ultraviolet (DUV) lithography, CAR has been optimized for extreme ultraviolet (EUV) applications to handle challenges related to resolution, sensitivity, and chemical uniformity [21]. Unlike SU-8, CARs leverage acid-catalyzed reactions to improve photosensitivity, reducing chemical blur, which contributes to improved line edge roughness (LER) [22]. Additionally, advances in CAR formulations have focused on minimizing chemical stochasticity and improving etch resistance, making them appropriate for innovative lithographic processes. Typically, CARs can reach resolutions of sub-14 nm [23] with the trade-off of lower aspect ratios around 5:1 [24].

SU-8 has been developed for over 20 years to create precise high-aspect-ratio microstructures, becoming a key material in MEMS and microfabrication. Its unique characteristics, such as increased mechanical strength, thermal stability, and chemical resistance, make it suitable for innovation and the creation of complex structures, such as microneedles, microfluidic filters, and cell sorters. Advanced lithography techniques have expanded their capabilities, including back-side exposure, inclined UV exposure, and holographic interference lithography. In parallel, CARs have gained importance in applications requiring higher resolution and improved etch selectivity, particularly in EUV lithography. Their high sensitivity and ability to achieve sub-14 nm features make them a key alternative in fabrication. Despite its maturity, SU-8 continues to evolve with optimizations for high-aspect-ratio, while CARs advance high-resolution lithography, guaranteeing that photoresists keep innovating in the microfabrication process, allowing improved performance and reliability in various applications.

7.3 Role of HSQ in High-Resolution, High-Aspect-Ratio Structures

HSQ has become a key material in high-resolution, high-aspect-ratio lithography, particularly in applications requiring nanoscale precision. As a negative tone resist, HSQ enables the fabrication of sub-10 nm structures while maintaining excellent mechanical and chemical stability [25]. Due to its high contrast and low line-edge roughness, it is widely used in electron beam lithography (EBL) and DUV lithography for advanced semiconductor manufacturing, photonics, and microelectromechanical systems (MEMS) [26].

Several studies have demonstrated the effectiveness of HSQ in achieving high-resolution patterning. Vila-Comamala et al. reported the successful fabrication of dense, high-aspect-ratio nanostructures using HSQ with 100 keV electron beam lithography. The study highlighted its exceptional etch resistance and pattern fidelity, which are crucial for applications in nano photonics and next-generation transistors [27]. Another study explored a bilayer resist stack using HSQ and polymethylmethacrylate (PMMA), demonstrating its ability to create complex three-dimensional nanostructures with improved process control [28]. These studies underscore HSQ's versatility in various high-resolution lithographic applications. However, despite these advantages, HSQ-based lithography still lacks sufficient comparative analysis with other resists like SU-8 and CAR.

HSQ's superior etch resistance has made it particularly valuable for nanoscale mask applications in semiconductor fabrication. Research indicates that it can withstand aggressive plasma etching, enabling precise pattern transfer onto silicon and dielectric substrates [26]. This characteristic is essential for defining ultra-fine features in integrated circuits, where pattern fidelity directly impacts device performance. Additionally, HSQ has been successfully utilized in optical waveguide fabrication, where its ability to form narrow, high-aspect-ratio features ensures efficient light propagation in photonic circuits [27].

Despite its advantages, HSQ presents several challenges in deep lithographic applications. One of the primary concerns is feature collapse, which occurs due to capillary forces during development and drying. Closely spaced nanostructures often adhere to one another, leading to pattern distortion and loss of structural integrity [27]. To resolve this issue, researchers have investigated alternative drying techniques such as critical point drying (CPD) and supercritical CO_2 drying, both of which help preserve high-aspect-ratio structures by eliminating surface tension effects.

Another challenge is HSQ's sensitivity to environmental factors, particularly its tendency to absorb moisture over time. This leads to degradation in pattern fidelity and mechanical stability, which can be detrimental to long-term device performance [25, 26]. Ongoing research is exploring protective coatings and modified formulations to enhance HSQ's resistance to humidity and oxidative degradation, ensuring long-term stability in nanofabricated structures. Recent advancements have led to the development of hybrid resist systems combining HSQ with organic polymers. These hybrid materials enhance processability while maintaining high resolution and etch resistance, making them particularly beneficial for EUV lithography and next-generation semiconductor manufacturing.

Optimizing the exposure and development processes is critical for maximizing HSQ's performance in high-resolution lithography. Since HSQ relies on a crosslinking mechanism upon exposure, precise control over the electron beam or DUV radiation dose is essential to achieve the desired resolution. Overexposure can cause excessive crosslinking, making development difficult, while underexposure can result in incomplete pattern formation. Researchers have developed advanced process strategies, including salty development techniques and optimized developer chemistries such as tetramethylammonium hydroxide (TMAH), to enhance HSQ's resolution and contrast [28].

To further improve HSQ's performance, hybrid resist systems that combine HSQ with organic polymers are being explored. These hybrid materials aim to enhance processability while maintaining high resolution and etch resistance, making them suitable for EUV lithography, which is expected to drive the next generation of semiconductor manufacturing [26].

Additionally, advances in thermal annealing and alternative crosslinking mechanisms are being investigated to improve HSQ's mechanical and chemical resilience, ensuring greater reliability in nanoscale applications [25, 28].

HSQ continues to play a critical role in high-resolution, high-aspect-ratio lithography, offering unmatched resolution, structural stability, and etch selectivity. Its application in semiconductor fabrication, photonics, and quantum technologies highlights its importance in next-generation nanotechnology. Despite existing challenges, ongoing research and process improvements are expanding HSQ's capabilities, ensuring its relevance in future advancements in meta-optics, quantum computing, and high-density electronic devices. Recent advancements have led to the development of hybrid resist systems combining HSQ with organic polymers. These hybrid materials enhance processability while maintaining high resolution and etch resistance, making them particularly beneficial for EUV lithography and next-generation semiconductor manufacturing.

Having discussed different types of photoresists, Table 7.1 summarizes each's key properties and achievable parameters. The values presented are representative, as different lithographic techniques, such as UV lithography, electron beam lithography, or variations in processing conditions like temperature, can significantly impact performance. Due to these variations, direct comparison of quantitative data is challenging, as each resist performs differently under optimized conditions.

7.4 Challenges in Deep Lithography

High aspect ratio lithography has several challenges that must be addressed with utmost care to realize the fabrication of high quality and reliable microstructures. Resistant swelling, loss of adhesion, and cracking are among the most significant challenges encountered in deep lithography and can lead to defects and structural

Table 7.1 Comparison between different photoresists

	SU-8	CAR	HSQ
Aspect-Ratio	20:1	5:1	10:1–15:1
Resolution	1 μm	14 nm	< 10 nm
Etch selectivity	Moderate	Very high	Very high
Young's module	4.02 GPa	3–5 GPa	40–60 GPa
Glass temperature	~50 °C	~100 °C	Not well defined
Degradation temperature	~380 °C	~400 °C	~ 300 °C
Thermal expansion coefficient	52 ± 5.1 ppm/K	Very low	Very low

failure. Other challenges like aspect ratio limitations and non-uniform exposure and development complicate the process. These challenges are controlled by a combination of resist formulation, exposure conditions, processing techniques, and environmental conditions, and optimized fabrication processes are needed to achieve accurate and reproducible results [29].

Swelling, caused by solvent intrusion into photoresist polymers while developing, distorts patterned features, particularly in thick resists such as SU-8. [30] Research showed that SU-8 swells in common developers (e.g., gamma-butyrolactone), at the expense of critical dimensions in HAR structures. Swelling occurs by solvent-driven polymer matrix plasticization, exacerbated by the long development time required for deep features [31].

Mitigation strategies for improved dimensional stability in deep lithography are primary strategies. Optimization of developers, e.g., via use of cyclopentanone-free developers, reduces contact between solvent and resist and consequently volumetric growth [32]. Also needed is control of post-exposure bake (PEB) processes. Multi-step PEB at 65–95 °C reduces residual solvent contents [33]. Another method is using alternative resists, i.e., CARs like poly methyl methacrylate (PMMA), which have less solvent affinity and are more dimensionally stable, further improving overall resist performance [4].

Adhesion failure between photoresists and substrates caused by residual stress or chemical incompatibility leads to delamination during wet etching or plasma processing. Also, SU-8 imposes more significant interfacial shear stresses on silicon substrates during cross-linking, causing delamination [34].

Surface priming techniques, such as depositing hexamethyldisilazane (HMDS), can form a hydrophobic monolayer on silicon dioxide that enhances adhesion [35]. In addition, depositing adhesion layers on non-silicon substrates, such as thin metallic interlayers like titanium, forms covalent bonds with resist polymers, enhancing adhesion strength [36]. Plasma treatment, specifically with oxygen plasma, increases surface roughness and facilitates mechanical interlocking and adhesion [37].

HAR cracking occurs due to thermal shrinkage tensile stresses during post-exposure baking or solvent evaporation. Also, during soft baking, evaporation of solvents leads to a reduction of volume and the creation of mechanical stress. With increasing film thickness and feature lateral dimensions, the stress grows. This leads to debonding of the resist layer from the substrate in case of weak adhesion. Recent studies have found that the largest contributor to total internal stress of the film during processing is soft-baking time, contributing up to 50%. Post-exposure bake time, development time, and exposure dose all contribute 30%, 15%, and 5%, respectively [4].

Stress relief methods are of crucial importance for maximizing the performance and life of lithographic processes. Gradual temperature ramping during soft baking, where the temperature is raised progressively at a controlled rate, is called graded soft baking. Graded soft baking helps in reducing thermal gradients, and therefore considerably lowers the stress in the resist film [4]. Solvent annealing is also extremely helpful, where slow evaporation of the solvent (for over 24 h at 25 °C) maintains capillary-induced stress at low levels during drying, and the resulting structure has increased integrity [38]. Research has proven that ultrasonic stress relief technology plays a significant role in the field of metal stress relief. The internal stress of metal material can be alleviated, and fatigue strength can be enhanced using this technology [32]. Additionally, hybrid lithography, an integration of laser ablation and UV lithography, helps to reduce localized stress concentrations in high-aspect-ratio (HAR) molds and decrease stress in the final structure even further [39].

7.5 Applications in MEMS and Other Microsystems

SU-8 is a general-purpose micromold material for electroplating applications. SU-8 is utilized in the production of high-aspect-ratio structures that serve as templates for electroplated metal microstructures, such as micro heat pipes. Etching the SU-8 away following electroplating can be employed in the fabrication of precise metallic parts required in MEMS applications [40] (Fig. 7.3).

Secondly, SU-8 is utilized as a master tool in hot-embossing processes that have been used in producing low-cost microstructures and more particularly BioMEMS

Fig. 7.3 Electroplating micro heat pipe using SU-8 as the molds. (Left)SU-8 molds; (Right) Cu electroplating microstructure [40]

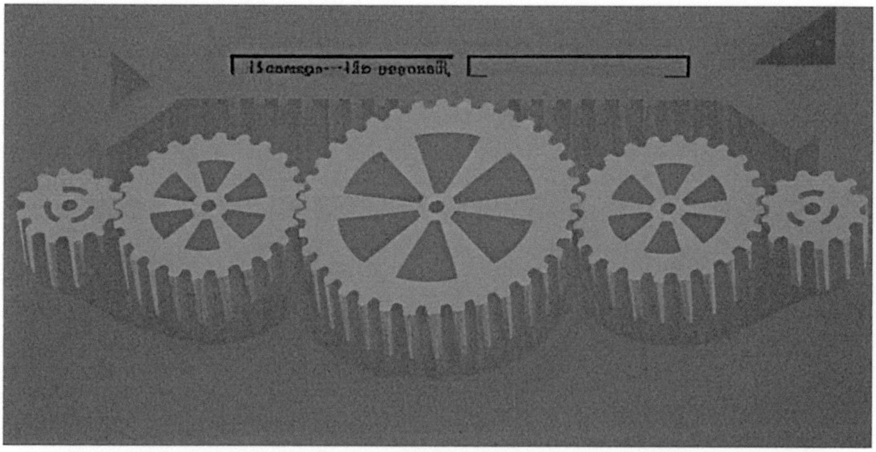

Fig. 7.4 Microstructure obtained by the optimized parameter with high-aspect-ratio and resolution [40]

products. The SU-8 master structure created is pressed into substrate materials such as polycarbonate or PMMA for microstructure replication, an important role fulfilled in disposable microfluidic chips production [40] (Fig. 7.4).

Since SU-8 is utilized as a structural material, complex microstructures such as embedded microchannels or microcavities can be formed entirely out of SU-8. This approach allows for the fabrication of integrated microsystems without additional bonding or layering steps [41]. SU-8 microstructures have even been successfully integrated into microcapillary electrophoresis (Micro-CE) systems, which utilize high-voltage separation techniques for analysis. SU-8 enables precise fabrication of microfluidic channels and integration with electrodes, which makes it an attractive material for lab-on-a-chip systems [40] (Fig. 7.5).

Due to its excellent adhesion and chemical resistance, SU-8 is applied as a mask material for wet etching. It is employed as a protection layer in substrate deep etching like glass, in order to obtain microscale patterns with high precision [40]. In addition, SU-8 plays a key role in creating micro-optical devices such as microlens arrays used for imaging and light manipulation, micro-resonators crucial for optical signal processing and photonic devices, and photonic crystals used in waveguides and integrated photonic circuits [41].

MPL in SU-8 has been widely used in the creation of photonic crystals (PCs) with NIR stopbands. It has also been used to create spatially variant photonic crystals (SVPCs) that can guide light through sharp bends. This is achieved by changing the orientation of the unit cells while keeping the lattice spacing and fill factor constant. SVPCs can curve light with a narrow radius of curvature, allowing beam size and profile flexibility. Compared to graded photonic crystals, SVPCs are inexpensive to fabricate. They also facilitate polarization control over light, as vertically polarized light curves while horizontally polarized light travels in a straight path [41] (Fig. 7.6).

Fig. 7.5 Microcapillary electrophoresis fabricated on glass substrate with SU-8 as the structural material and Pt as the electrode [40]

In biomedical systems, SU-8 is used in the fabrication of Nano and microfluidic devices for biosensing, drug delivery, and lab-on-a-chip systems. These devices utilize the high biocompatibility and mechanical stability of SU-8, making it a material of choice for biomedical engineering [41] (Figs. 7.7, 7.8).

Furthermore, SU-8 is used to create high-aspect-ratio elastic structures required for MEMS actuators, including piezoelectric motors and compliant microstructures. These components are essential in precision mechanical systems where miniaturization and mechanical robustness are criticalm [40]. High-aspect-ratio SU-8 lithography has enabled significant advancements in MEMS and microsystem technologies. Its application ranges from microfluidics, electroplating, and optics to biomedical devices and mechanical MEMS, indicating its potential as a basic material for microfabrication [40, 41].

Recent advancements in high aspect ratio lithography (HARL) have expanded its use beyond research laboratories into high-volume manufacturing and commercial semiconductor fabrication [26]. The following subsections outline process innovations, material improvements, and case studies that demonstrate the transition of HARL to industrial production.

Advanced lithography hardware such as automatic step-and-repeat systems coupled with DUV, and EUV exposure sources have radically enhanced throughput without the loss of critical dimension control [26]. Process optimization, such as more advanced post-exposure and development techniques, has reduced thermal and mechanical processing stresses, resulting in higher yield and more uniform performance [33]. Improved SU-8 formulations now offer improved mechanical stability, reduced swelling, and better resolution uniformity, which are essential for high-yield

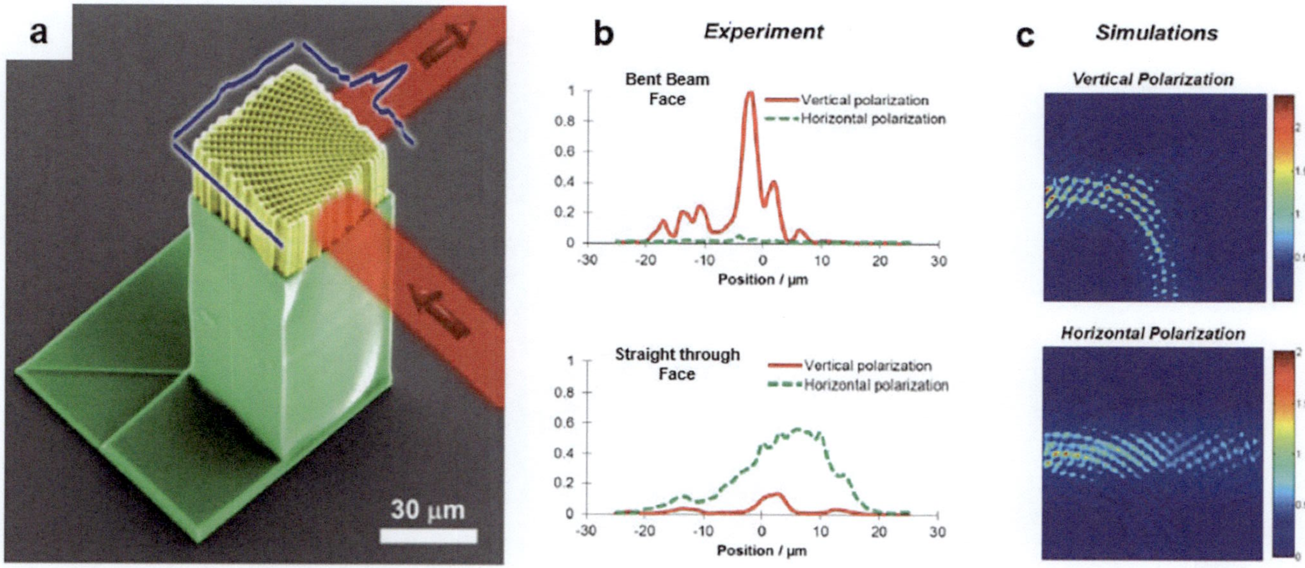

Fig. 7.6 Spatially variant photonic crystals (SVPCs) fabricated by MPL in SU-8. **a** False-color SEM image of the SVPC (shown in yellow). The blue ribbons depict experimentally measured relative intensity exiting each face of the SVPC when an optic fiber introduced light having $\lambda_o = 2.94\ \mu$ m onto the lattice. **b** Polarization dependence of beam bending-for the same SVPC. **b** Experimentally measured intensity for vertically or horizontally polarized light exiting at faces corresponding to straight-through transmission and beam-bending. **c** Simulations of beam bending in the SVPC [42]

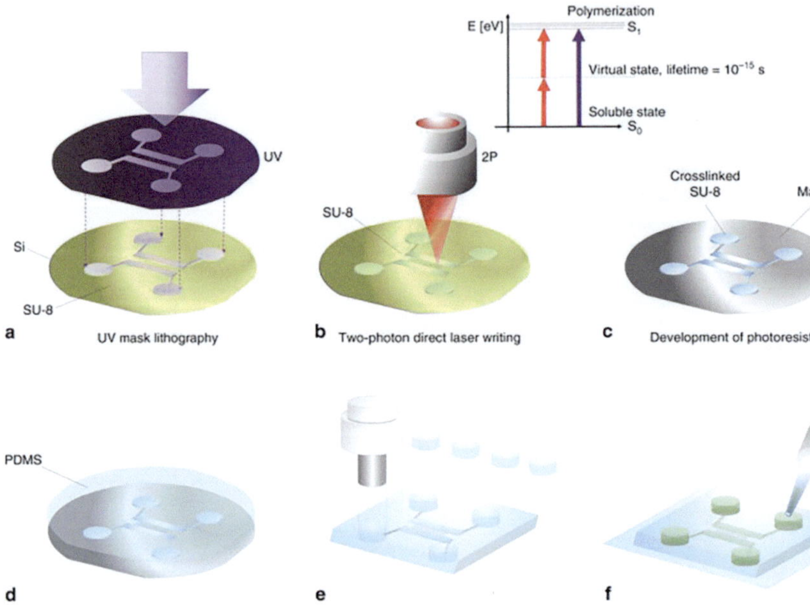

Fig. 7.7 Process outline for nanofluidic device fabrication via combination of two-photon lithography and mask UV lithography, **a** Project an arbitrary microfluidic chip design onto a SU-8 coated silicon wafer using mask-based UV lithography. **b** Add nanofluidic channels in specific areas of interest using two-photon lithography. **c** Develop the wafer to obtain a microfluidic master wafer with integrated nanofluidics **d** Coat the wafer with PDMS and cure it to form the device **e** Peel the cured PDMS device from the wafer and add inlets **f** Plasma bond the final PDMS chip onto a glass coverslip and fill it with a solution, e.g., fluorescent dye [43].

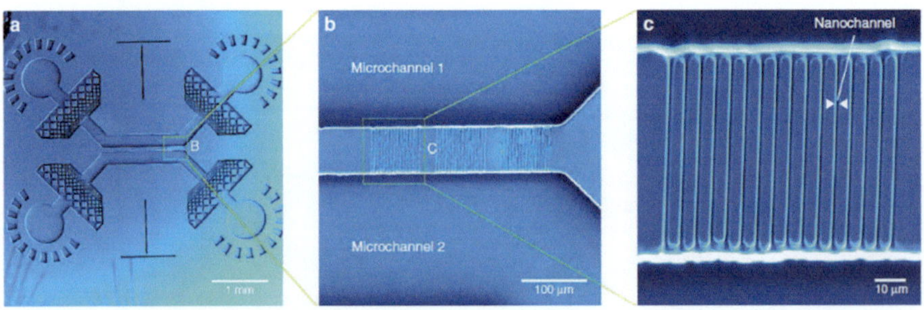

Fig. 7.8 SEM of the nano-fluidic device fabricated by photolithography and MPL. **a** Device after PDMS molding. **b** 75-m nano-channels connecting the two micro-channels. **c** Magnified view showing the 420-nm wide nano-channels [43]

manufacturing [4]. HSQ development enables sub-10 nm high-resolution patterning that meets the precision demands of both research and industrial applications [26].

HAR lithography is applied in the production of high-volume MEMS sensors, micro heat pipes, and microfluidic chips, where precise high-aspect-ratio structures are critical for reliable performance [12]. The technique is also becoming the prime enabler in manufacturing three-dimensional memory chips such as 3D NAND flash memory, whose vertical stacking of cells relies on reliable high-AR patterning for achieving storage density increases and improved performance of devices [26].

Commercial semiconductor fabs are incorporating HAR lithography to produce the ultra-fine features required for next-generation devices, including through-silicon vias, advanced interconnects, and complex three-dimensional architectures [25]. Advanced process control techniques, such as enhanced metrology and exposure monitoring, ensure that strict tolerances for pattern fidelity and uniform development are met in high-volume production environments [3, 11]. Although HARL offers benefits such as increased integration density and improved device performance, problems such as resist swelling and process sensitivity continue to plague it and are being addressed by current research into high-aspect-ratio patterning with chemically amplified resists and stress reduction methods [31, 33].

High-Aspect-Ratio Lithography (HARL) has proven to be a transformative technology in microsystems fabrication, enabling the development of three-dimensional structures essential for applications in MEMS, microfluidics, and photonics. The advancements in SU-8, HSQ, and chemically amplified resists have expanded the resolution and aspect ratios achievable, significantly impacting device miniaturization and performance. The ongoing research in resist formulation, exposure techniques, and post-processing strategies continues to improve HARL results, addressing challenges such as resist swelling, adhesion loss, and structural cracking.

The increasing demand for high-resolution lithographic processes has driven HARL's integration into industrial manufacturing, particularly in semiconductor fabrication, 3D NAND memory, and MEMS sensor production. Innovations such as EUV lithography and hybrid resist systems are expected to enhance HARL's precision and scalability, making it essential for new microfabrication technologies.

References

1. R.M.M. Hasan, X. Luo, Nanomanufacturing and Metrology **1**, 67–81 (2018)
2. K. Kamranikia, S. Dominici, M. Keller, N. Kube, K. Mougin, A. Spangenberg, *Micromachines (Basel)* 14, (2023)
3. A.L. Bogdanov, S.S. Peredkov, Microelectron. Eng. **53**, 493–496 (2000)
4. A.d. Campo, C. Greiner, *Journal of Micromechanics and Microengineering* 17, R81-R95 (2007)
5. M. Sebald, *Chimia* 47, (1993)
6. J.M. Warlaumont, G.N. Taylor, R.S. Hutton, S.M. Stein, C.H. Boyce, O.R. Wood Ii, B. LaFontaine, A.A. MacDowell, D.R. Wheeler, G.D. Kubiak, A.K. Ray-Chaudhuri, K.W. Berger, D.A. Tichenor, Extreme UV resist technology: the limits of silylated resist resolution, Electron-Beam, X-Ray, EUV, and Ion-Beam Submicrometer Lithographies for Manufacturing V, (1995)

7. J.M. Park, J.H. Kim, J.S. Han, D.S. Shin, S.C. Park, S.H. Son, S.J. Park, *Materials (Basel)* 12, (2019)
8. G.W. Reynolds, J.W. Taylor, Journal of vacuum science & technology B: microelectronics and nanometer structures processing. Measurement, and Phenomena **17**, 334–344 (1999)
9. B. Wu, A. Kumar, S. Pamarthy, *Journal of Applied Physics* 108, (2010)
10. H.J. Levinson, Y. Ma, M. Koenig, B. La Fontaine, R. Seltmann, Proposal for determining exposure latitude requirements, Optical Microlithography XXI, (2008)
11. T. Christenson, *Journal of Micro/Nanolithography, MEMS, and MOEMS* 3, (2004)
12. H. Lorenz, M. Despont, N. Fahrni, J. Brugger, P. Vettiger, P. Renaud, Sens. Actuators, A **64**, 33–39 (1998)
13. X.L. Zhang, L.Q. Du, A.A. Wang, Key Eng. Mater. **562–565**, 96–101 (2013)
14. G. Voskerician, M.S. Shive, R.S. Shawgo, H. von Recum, J.M. Anderson, M.J. Cima, R. Langer, Biomaterials **24**, 1959–1967 (2003)
15. H. Shao, X. Hang, D. Jiang, *Physics Letters A* 523, (2024)
16. M.R. Haq, H. Schift, *Micro and Nano Engineering* 19, (2023)
17. M. Aktary, M.O. Jensen, K.L. Westra, M.J. Brett, M.R. Freeman, Journal of vacuum science & technology b: microelectronics and nanometer structures processing. Measurement, and Phenomena **21**, L5–L7 (2003)
18. Y.J. Chuang, F.G. Tseng, W.K. Lin, Microsyst. Technol. **8**, 308–313 (2002)
19. J. Lee, K.-H. Choi, K. Yoo, Micromachines **6**, 1–18 (2014)
20. K. Lee, H.C. Lee, D.S. Lee, H. Jung, Adv. Mater. **22**, 483–486 (2010)
21. J. Cen, Z. Deng, S. Liu, Polym. Chem. **15**, 4599–4614 (2024)
22. M. Hassaan, U. Saleem, A. Singh, A.J. Haque, K. Wang, *Materials (Basel)* 17, (2024)
23. T. Kimoto, T. Naruoka, H. Nakagawa, T. Fujisawa, M. Shiratani, T. Nagai, R. Ayothi, Y. Hishiro, M. Hori, K. Hoshiko, T. Kimura, J. Photopolym. Sci. Technol. **28**, 519–523 (2015)
24. S. Pauliac-Vaujour, P. Brianceau, C. Comboroure, O. Faynot, Microelectron. Eng. **85**, 800–804 (2008)
25. Q. Lin, A.E. Grigorescu, M.C. van der Krogt, C.W. Hagen, Sub-10-nm structures written in ultra-thin HSQ resist layers using electron-beam lithography, Advances in Resist Materials and Processing Technology XXIV, (2007)
26. A.E. Grigorescu, C.W. Hagen, Nanotechnology **20**, 292001 (2009)
27. J. Vila-Comamala, S. Gorelick, V.A. Guzenko, E. Farm, M. Ritala, C. David, Nanotechnology **21**, 285305 (2010)
28. H.W. Do, J.-B. Chang, K.K. Berggren, *Journal of Vacuum Science & Technology B, Nanotechnology and Microelectronics: Materials, Processing, Measurement, and Phenomena* 32, (2014)
29. K.R. Williams, R.S. Muller, J. Microelectromech. Syst. **5**, 256–269 (1996)
30. H. Lorenz, M. Despont, N. Fahrni, N. LaBianca, P. Renaud, P. Vettiger, J. Micromech. Microeng. **7**, 121–124 (1997)
31. M. Chatzichristidi, E. Valamontes, P. Argitis, I. Raptis, J.A. van Kan, F. Zhang, F. Watt, Microelectron. Eng. **85**, 945–948 (2008)
32. N.C. LaBianca, J.D. Gelorme, Proceedings of Spie the International Society for Optical **2438**, 846–852 (1995)
33. X. Qu, J. Li, Z. Yin, Microsyst. Technol. **24**, 3141–3147 (2018)
34. G. Barillaro, L.M. Strambini, V. Annovazzi-Lodi, S. Merlo, IEEE J. Sel. Top. Quantum Electron. **15**, 1359–1367 (2009)
35. K.E. Petersen, Proc. IEEE **70**, 420–457 (1982)
36. K.D. Wise, D.J. Anderson, J.F. Hetke, D.R. Kipke, K. Najafi, Proc. IEEE **92**, 76–97 (2004)
37. H.V. Jansen, M.J. de Boer, S. Unnikrishnan, M.C. Louwerse, M.C. Elwenspoek, *Journal of Micromechanics and Microengineering* 19, (2009)
38. C.G. Khan Malek, *Anal Bioanal Chem* 385, 1362–9 (2006)
39. J. Arai, H. Kawai, F. Okano, Appl. Opt. **45**, 9066–9078 (2006)
40. J. Zhang, K.L. Tan, H.Q. Gong, Polym. Testing **20**, 693–701 (2001)
41. P. Golvari, S.M. Kuebler, *Micromachines (Basel)* 12, (2021)

42. J.L. Digaum, J.J. Pazos, J. Chiles, J. D'Archangel, G. Padilla, A. Tatulian, R.C. Rumpf, S. Fathpour, G.D. Boreman, S.M. Kuebler, Opt. Express **22**, 25788–25804 (2014)
43. O. Vanderpoorten, Q. Peter, P.K. Challa, U.F. Keyser, J. Baumberg, C.F. Kaminski, T.P.J. Knowles, Microsyst. Nanoeng. **5**, 40 (2019)

Chapter 8
Challenges in Photoresist Technology for Microsystems

8.1 Materials and Processing Limitations

Photoresist is a light-sensitive material used in fabrication of microsystems, especially in photolithography procedures. It is used to create precise patterns on wafer substrates which are considered the backbone of microelectronics manufacturing [1]. Photoresists have crucial parameters, such as resolution; sensitivity; stability and scalability, and these parameters are important in terms of designing accurate wafer substrates to create functional microsystems. These parameters are mainly affected by the processing and the inner properties of the photoresist. In this part, we will discuss how these major points can create challenges for the making of photoresists, and how to overcome such issues.

Resolution: The chemical properties of the photoresist such as its composition, thickness and molecule structure, influence how the resist image will be formed. For example, chemically amplified photoresists use photoacid generator (PAG) which produce acid to trigger specific reactions. If the acid still reacts for longer than the specified time, the image of the film would blur, and this phenomenon is known as acid diffusion. Furthermore, the need to maintain thin films to minimize light scattering and improve contrast will be on the cost of resolution restriction [2].

On the other hand, processing parameters such as exposure dose; bake temperature and time; and the development conditions, are also considered as major factors from the perspective of processing. Over-baking or under-baking can lead to incomplete reactions and excessive acid diffusion, which both affect the resolution of the photoresist [2].

For the case of electron beam lithography, electron scattering within the resist material can limit the resolution. Lift-off process using chemically amplified (CAMP) negative photoresist has its own challenges. The lift-off process can and does show multiple types of defects and weaknesses, these defects mainly can occur in three segments, photolithography; metal deposition or resist removal [3]. When creating

K. Wang, *Photoresist Technology in Microsystems: Principles, Processes and Applications*, Springer Series in Advanced Microelectronics 15,
https://doi.org/10.1007/978-981-95-1606-3_8

high density features, overall accuracy will be reduced due to the interaction between electron beam and the photoresist [4].

Stability: Chemical resistance and adhesion are related to the long-term stability of a photoresist. Materials with poor adhesion tend to produce patterns that have inconsistent dimensions and degraded sidewall profiles. On the other hand, using advanced resists can improve stability, but that will require more precision and control during the processing, hence more complicated fabrication steps [5].

While stress during coating, baking and development can lead to major defects like cracking or pattern collapse. That will affect both pattern fidelity and resist's performance during later steps [5].

Furthermore, the compatibility of photoresist materials with wafer substrates and processing conditions may affect overall stability. For example, uneven coating and improper curing may lead to material contamination, which also can lead to several defects like unintended exposure [6].

Scalability: Scaling the photoresist films to smaller dimensions like 50 nm or 20 nm will be accompanied by challenges, and new considerations will occur, such as the length of acid diffusion or the molecular weight of the polymer. For instance, using higher concentrations of PAGs might improve sensitivity, but also it might increase diffusion-related blurring, which could limit the resist patterning [7]. In processing steps, uniformity over the wafer area becomes very challenging to maintain at very small dimensions. The narrower the window that is needed to be achieved, the more sensitive the process would become. Using Extreme Ultra-Violet photoresist (EUV) materials, which are light sensitive to EUV photons, would generate photoacids, and this reaction enhances sensitivity. But this sensitivity can limit reproducibility, which is considered as an important point to achieve scalable manufacturing [7].

In addition, smaller dimensions will lead to more complex lithography procedures, this will affect scalability especially when dealing with advanced nodes and 3D structures, such as DRAM manufacturing. Choosing the right methodology in this case is crucial. Metal Oxide Resist (MOR) has better line-edge roughness (LER) and defectivity at lower exposure, despite the low thickness in comparison with Chemically Amplified Resist (CAR) [8].

From Table 8.1, we can see that CARs are cost-effective and scalable because of their long-established presence in the industry. On the other hand, CARs face some limitations in achieving ultra-fine resolutions for upcoming technology nodes in the future, not like MORs, which offers superior resolution potential, the issue with MOR technology that it is still under development [9, 10].

Proposed Solutions: These challenges can be overcome by utilizing several techniques which will be mentioned shortly, bear in mind that we should also consider features' trade-off when applying some of these solutions.

Utilizing hybrid photoresist systems may serve several key advantages. These hybrid photoresists are composed of organic and inorganic materials, which are involved with a technique called Vapor-Phase Infiltration (VPI). This method allows precise control over composition properties of the hybrid material [11]. These materials enable high-resolution patterning, the ability to achieve 50 nm diameter hole patterns (Fig. 8.1).

Table 8.1 Comparison between MOR and CAR photoresists

	Chemically amplified resists (CARs)	Metal oxide resists (MORs)
Cost	Lower	Higher
Resolution	Limited (>25nm tip-to-tip)	Superior (down to 23nm tip-to-tip)
Scalability	Challenges beyond 28nm pitch	Better potential for scaling
Process Window	Narrower	Wider
Ecosystem	Well-established	Developing

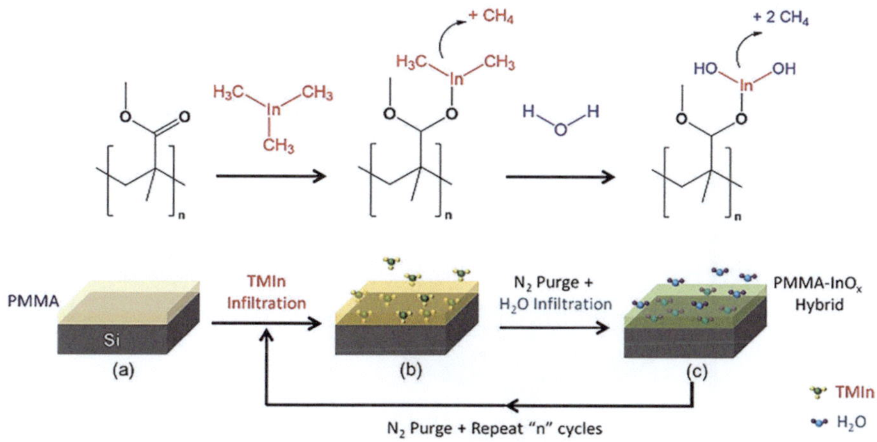

Fig. 8.1 The VPI synthesis of PMMA-InO_x hybrid resist [11]

Since we noticed that some factors are affected by processing control, Enhanced monitoring steps are crucial in terms of improving photoresist quality. In-Situ monitoring of photoresist thickness contour on wafers is a novel method to achieve processing control. Monitoring conducted right after the spin-coat or edge-bead removal. A spectrometer is utilized in real-time to allow immediate inspection of the wafer thickness. This is very helpful in terms in early detection of defective wafers and prevention of unnecessary and excessive costs [12] (Fig. 8.2).

Multi-Point inspection strategy is another method to improve processing quality. It helps controlling defect formation by conducting multiple inspection points; it also helps improve the correlation between inspection results and electrical measurements [13].

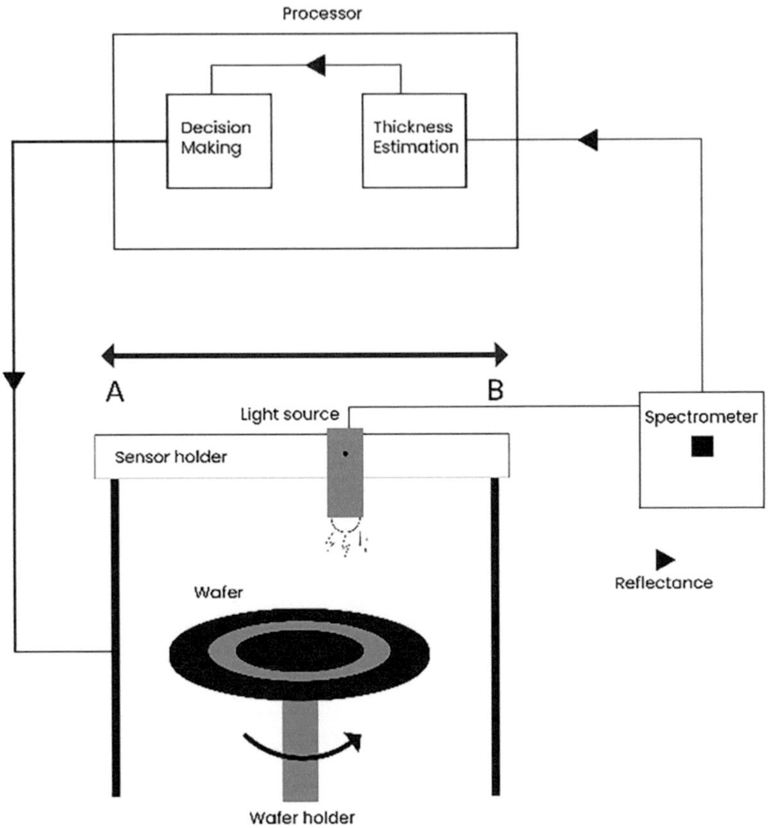

Fig. 8.2 Experimental setup shows real-time monitoring of wafer thickness

8.2 Scaling Down: Challenges at the Nanoscale

Refining lithographic processes to achieve sub-20 nm patterning requires not only advanced instrumentation but also careful control of material formulations and exposure/development parameters. Ultrafine features make both quantum effects and molecular interactions necessary to be considered, since even minute chemical or morphological variations can jeopardize device fidelity. Through innovations in non-chemically amplified resists (n-CARs), photoinitiators, and stimulus-responsive polymers, researchers aim to push nanoscale resolution further—while preserving key properties such as photoresist stability, sensitivity, and biocompatibility.

Impact of Quantum Effects: At the nanoscale, quantum confinement becomes a key driver of material properties. In colloidal quantum dots (QDs), for instance, reducing particle size below the exciton Bohr radius causes discrete energy levels to form, profoundly altering electronic and optical behavior compared to bulk materials which leads to size-tunable photoluminescence and high color purity, making QDs appealing for displays, sensors, and other optoelectronic devices [14]. However, these same quantum-scale phenomena also increase sensitivity to processing conditions.

Because QDs rely on molecular ligands for stability, their solubility can complicate standard top-down lithography: typical photoresists or solvents can dissolve or damage QD films. Quantum-level uniformity becomes harder to maintain when employing solution-based deposition, as minor variations in surface ligands or environment can shift the QD energy levels. As device features approach quantum-dot dimensions, any inhomogeneity in QD size or local environment leads to spectral broadening and reduced performance. Achieving high-resolution patterning (e.g., tens of nanometers) is critical to preserve quantum effects—otherwise, inter-dot coupling can alter emission or transport properties.

QDs have large surface-to-volume ratios (Fig. 8.3), so their emission and electrical characteristics are heavily influenced by interface chemistry or nearby layers. Dry processes, such as the photoresist-contact approach helps preserve QD integrity by avoiding harsh solvents that disrupt quantum confinement and degrade the ligands [14]. Overall, quantum confinement in semiconducting nanoparticles mandates precise, gentle lithographic methods. Even small chemical or morphological variations may shift the QD bandgap or fluorescence intensity, emphasizing the importance of careful process control at the nanoscale.

Impact of Molecular Interactions: When critical feature sizes drop below 20 nm, molecular-level interactions within photoresists become increasingly consequential for final pattern fidelity and process reliability. In non-chemically amplified resists (n-CARs), these interactions are especially crucial because the primary mechanism of solubility switching—whether chain scission, crosslinking, or photodeprotection—must occur within a tight molecular environment, without external acid diffusion (as in chemically amplified resists).

Many n-CARs are polymeric, and their coil-like backbones can entangle heavily. Below 20 nm, these entanglements influence dissolution rate and can manifest as line-edge roughness (LER) if the polymer network dissolves unevenly. Polymers or small molecules with strong hydrogen-bonding sites or aromatic stacking can form aggregated domains in the solid film. During development, these aggregates dissolve more slowly or unevenly, potentially increasing pattern roughness. Amorphous materials typically yield more uniform patterning; if the n-CAR (non-chemically amplified resists) crystallizes (e.g., certain small-molecule or organometallic resists), grain boundaries or incomplete crystallites induce rough edges.

Incorporation of metal centers or inorganic clusters can increase absorption cross-section, improving sensitivity and etch resistance. However, phase separation or poor compatibility at organic–inorganic interfaces can degrade resolution and cause rough features. Post-exposure bake (PEB) or other thermal steps allow chain relaxation, impacting dissolution contrast in the exposed vs. unexposed areas. Minor variations in free-volume or polymer mobility at sub-20 nm can profoundly affect final feature sizes and uniformity. Choice of polymer architecture (e.g., comb polymers vs. linear chains) can help reduce chain entanglements, thereby minimizing LER in sub-20 nm patterns. Controlling hydrogen-bonding or aromatic packing—for instance through functional group selection—helps produce more uniform solubility

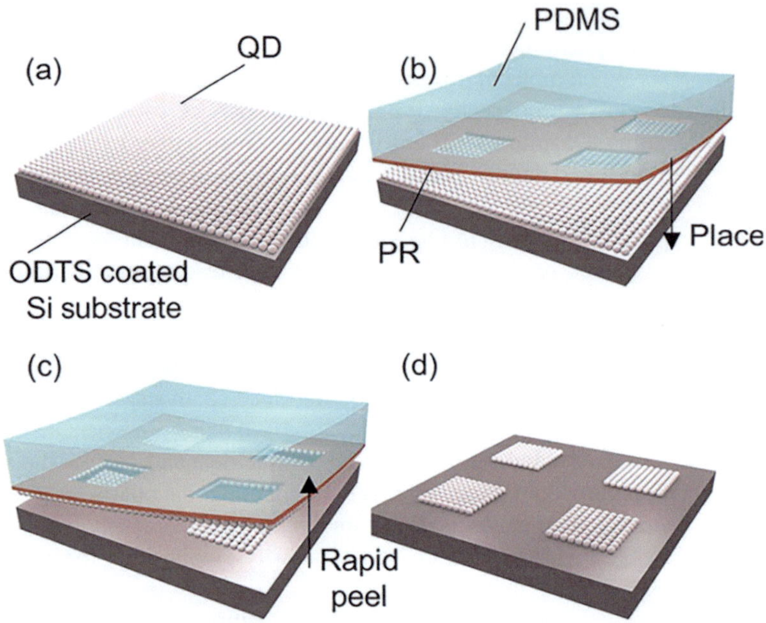

Fig. 8.3 Schematic process flow and images for PR contact patterning of a QD film on an ODTS-coated Si substrate. **a** A QD film is spincast on an ODTS-coated Si substrate. **b** The separately prepared PR stamp, i.e., a PR layer patterned on a PDMS substrate, is brought into contact with the QD film. **c** Subsequent peeling of the PR stamp results in a patterned QD film on the Si substrate. **d** Patterned QD film on the Si substrate [14]

changes and edges. Amorphous molecular resists can avoid grain-boundary roughness, yielding sharper sidewalls. Metallic or inorganic species (e.g., HFO_2, polyoxometalates, tin–oxo clusters) enhance sensitivity and contrast, but require careful tailoring to prevent phaseseparation or inadequate film uniformity. Ultimately, by refining these molecular interactions, n-CARs can achieve both high resolution and low roughness in next-generation lithography, addressing the demands of sub-20 nm node device fabrication [15].

Additionally, process parameters are crucial to lithographic pattern fidelity in the micro/nano regime, especially exposure time and development time for positive-tone photoresists. Controlling exposure intensity, development duration, and baking conditions is essential for consistent pattern transfer at nanometer scale [16]. By combining these insights with advanced n-CAR formulations, researchers can further push the resolution, reliability, and throughput of next-generation lithographic processes. TPP lithography's reliance on ultrafast laser pulses—combined with the right selection of photoinitiators—can mitigate these issues by shrinking the effective voxel size and curtailing undesired scattering or local overheating. Adjusting the laser scan speed and dose can systematically balance resolution and throughput in large-scale, complex biomedical devices [17].

Recent research has demonstrated that submicrometer resolution in photoresists is achievable by developing nano-interpenetrating semiconducting polymers, which combine high resolution with robust electrical performance even after repeated solution-based processing. At nanoscale feature sizes, conventional polymer resists often face metastable molecular arrangements: Cross-linked side chains or harsh development steps can disrupt the π–π stacking essential for charge transport. This disruption is exacerbated by shot noise during exposure (especially in electron-beam and deep-UV lithography), which creates stochastic variations in critical dimensions [18].

Notably, side-chain cross-linking within the bulk of conjugated polymers may disturb coherent charge transport, resulting in lower mobilities and device instabilities. By contrast, confining cross-linking networks within interpenetrating structures can stabilize π–π stacking, improving both mechanical and chemical resilience at sub-50 nm scales. Such architectures reduce pattern collapse and preserve charge transport properties against repeated soaking in organic solvents (i.e., developers and strippers). Consequently, when scaling down to the nanoscale, material design must address both resolution limits and molecular packing, ensuring that quantum-level variations and capillary forces do not degrade device fidelity. This approach paves the way for all-photolithography organic electronics with high densities and stable performance [18].

8.3 Environmental and Health Concerns

Environmental Impact of Photoresist Technology: The processing of photoresists in semiconductor fabrication is a major source of environmental pollution, particularly through the release of volatile organic compounds (VOCs) such as benzene, toluene, xylene, and PGMEA. These chemicals increase airborne contamination in semiconductor cleanrooms, raising occupational health concerns. Additionally, by-products like phenol and cresol can contaminate water sources, leading to long-term environmental damage if not effectively managed. Studies indicate that VOC emissions from photoresist processing can reach up to 150 ppm in cleanroom environments, significantly exceeding OSHA's recommended exposure limit of 50 ppm for prolonged exposure. These risks highlight the urgent need for better solvent control and waste management in semiconductor facilities [19].

Lithography, the most chemical-intensive step in semiconductor manufacturing, significantly contributes to environmental hazards. Photoresist solvents such as PGMEA, ethyl acetate, and acetone are among the most widely used chemicals in the industry. If not properly disposed of, these substances can contaminate air and water systems, increasing environmental toxicity. Once exposed to UV light, photoresists break down into harmful compounds like benzene, toluene, and xylene, which can persist in the environment and affect ecosystem stability. Additionally, the thermal decomposition of photoresists generates hazardous by-products such as formaldehyde, further complicating waste disposal challenges [20].

Extreme ultraviolet (EUV) lithography, the dominant technique for advanced semiconductor fabrication, presents another set of environmental concerns. The process relies on laser-generated plasma, which has low energy conversion efficiency, resulting in high power consumption. As semiconductor technology advances, energy usage in fabs has increased by nearly 60% from 7 to 3 nm technology nodes, making lithography one of the most energy-intensive steps in chip production. To address this, researchers are developing free-electron lasers as an alternative high-efficiency EUV light source, which could significantly lower power consumption [21].

Another challenge stems from spray-coated photoresists, which tend to form dry resist particles due to solvent evaporation. These dry droplets, if not controlled, lead to wafer contamination, increased material waste, and higher defect rates in semiconductor processing. Poor solvent vaporization control results in significant chemical loss, accumulating as hazardous waste within fabrication plants [22].

Health Hazards Associated with Photoresist Chemicals: Many chemicals used in photoresist formulations pose significant health risks, particularly for workers in photolithography areas. Benzene exposure has been linked to leukemia and blood disorders, while toluene and xylene affect cognitive function, leading to neurological damage, dizziness, and memory impairment. According to OSHA guidelines, the permissible exposure limits (PELs) for benzene, toluene, and xylene are 1 ppm, 200 ppm, and 100 ppm, respectively. However, workplace air monitoring has shown that benzene levels in semiconductor fabs frequently exceed this limit, sometimes reaching up to 5 ppm in poorly ventilated areas [19].

PGMEA vapors are associated with respiratory irritation and liver toxicity, while phenol and cresol may lead to immune system disorders and skin irritation. Due to frequent exposure, workers face long-term health risks. A Bayesian risk assessment model has been used to evaluate worker exposure to photoresist chemicals. Figure 8.4 illustrates hazard quotient (HQ) values for benzene, phenol, PGMEA, and xylene, showing that exposure levels often exceed safety limits, emphasizing the need for stricter safety controls [19].

Certain photoresists contain carcinogens like cyclohexanone, ethylbenzene, and pyridine, increasing risks of neurological disorders and reproductive toxicity [20]. Table 8.2 lists common carcinogenic compounds found in photoresists, highlighting discrepancies in MSDS disclosures, suggesting an urgent need for regulatory enforcement. Stronger ventilation, PPE, and air quality monitoring are necessary to minimize worker exposure and ensure safer semiconductor fabrication environments.

Regulations on Chemical Emissions and Waste Disposal: Although benzene regulations exist, its presence in photoresist by-products is frequently overlooked, creating gaps in semiconductor industry safety compliance. Toluene and xylene are subject to occupational exposure limits, yet enforcement remains inconsistent across semiconductor facilities. Similarly, PGMEA, despite being widely used, lacks comprehensive regulatory control, increasing worker exposure risks [19].

There are currently no standardized global regulations specifically targeting photoresist solvent emissions and hazardous waste disposal. Agencies like the Environmental Protection Agency (EPA) and the European Chemicals Agency (ECHA)

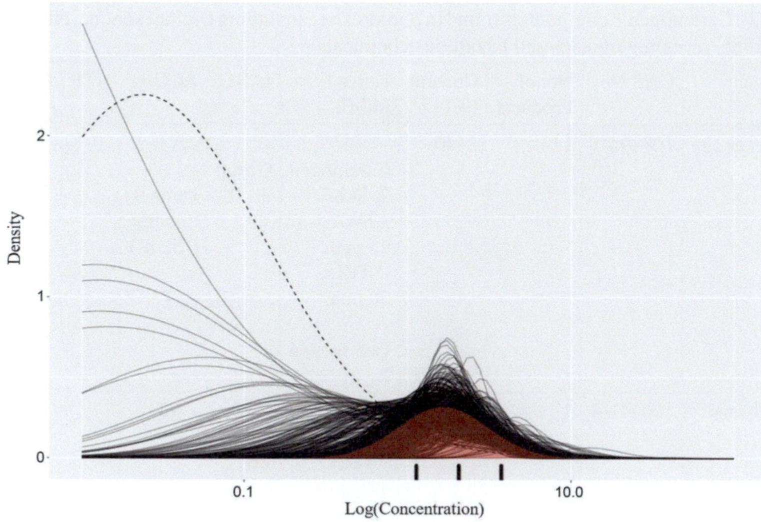

Fig. 8.4 Bayesian model showing hazard quotients (HQs) for benzene, phenol, PGMEA, and xylene, identifying excessive exposure risks [19]

have established restrictions on benzene and ethylbenzene, but enforcement varies significantly between regions and across different semiconductor manufacturers [17].

Another major concern is the accuracy of Material Safety Data Sheets (MSDS) for photoresist products. Studies have found that some manufacturers fail to disclose hazardous ingredients such as 1,4-dioxane and 2-butoxyethanol, despite their known toxicity. While some nations impose specific restrictions on benzene and toluene emissions, there is no globally accepted regulatory framework to control solvent emissions from photoresist processing. This highlights the urgent need for stricter chemical disclosure policies and stronger regulatory oversight in semiconductor manufacturing [18].

Sustainable Alternatives and Future Improvements: Reducing the environmental and health risks associated with photoresist technology requires a multi-faceted approach. Enhancing air filtration systems, enforcing protective equipment (PPE) standards, and conducting regular air quality assessments are essential steps in minimizing worker exposure. Additionally, the development of water-based photoresists and the implementation of solvent recycling systems present promising solutions for lowering chemical emissions and improving workplace safety [19].

To improve EUV lithography efficiency, researchers are exploring free-electron lasers, which could significantly reduce electricity consumption. Additionally, alternative lithographic techniques, such as nanoimprint lithography (NIL), directed self-assembly (DSA), and multi-electron beam lithography, are being investigated to decrease material waste and enhance production efficiency. Advances in computational lithography are also helping optimize exposure doses and lower chemical consumption [21].

Table 8.2 Carcinogenic compounds found in photoresists, including cyclohexanone, ethylbenzene, and pyridine, revealing undisclosed hazardous chemicals

Constituent	CAS No	No. of Products	Content (%)	Korea MOEL	IARC	ACGIH	NTP	EU CLP
Cyclohexanone	108–94-1	11	3–40	- Carcinogen 2, Skin - TWA: 25 ppm -STEL: 50 ppm - Carcinogen 2	- Group 3	- A3 - Dermal toxicity ('Skin')		
Ethylbenzene	100–41-4	1	0.1–1.0	-TWA: 100 ppm - STEL: 125 ppm	- Group 2B	- A3		
Pyridine	110–86-1	1	0.1–1.0	- Carcinogen 2 - TWA: 2 ppm	- Group 3	- A3		
1,4-Dioxane	123–91-1	4	< 1.0	- Carcinogen 2 - Dermal toxicity ('Skin') - TWA: 20 ppm	- Group 2B	- A3 - Dermal toxicity ('Skin')	- R	- Carcinogen 2

New developments, including block copolymer (BCP) lithography and polymer brush lithography, offer the potential to eliminate toxic photoresists, thus reducing emissions and waste production. Introducing solvent recovery systems in fabs could further reduce PGMEA and ethyl acetate emissions, lowering both air pollution and hazardous waste buildup [17].

In terms of performance, lacquer-based photoresists exhibit excellent properties such as high hardness (0.78–0.89 relative to glass transition value), electrical breakdown strength (50–80 kV/mm), and thermal stability (up to 250 °C short-term and 150 °C long-term), making them suitable for microelectronics applications. While the initial extraction and processing costs of lacquer are high, its long-term benefits, such as reduced environmental pollution and improved durability, could offset these costs. The paper suggests that further innovations in cost-effective processing methods could enhance the economic viability of lacquer-based photoresists, making them a promising green alternative in the semiconductor industry [23].

Another promising approach involves developing photoresist formulations that do not rely on hazardous solvents like PGMEA and cyclohexanone, thereby reducing exposure risks and environmental concerns. Additionally, acetone vapor treatment has been identified as an effective method for removing dry resist residues, improving pattern quality while reducing contamination. Controlling nitrogen flow in spray-coating systems can further prevent premature solvent evaporation, ensuring more uniform resist coverage and less material waste. Lastly, AZ4999 photoresist has demonstrated strong corrosion resistance, significantly lowering the corrosion rate of CoFe magnetic films, making it a viable option for improved sustainability in semiconductor applications [22].

8.4 Addressing Defects and Process Variability

Microelectromechanical systems (MEMS) and other microsystem technologies rely heavily on photolithography, a process where photoresists play a critical role in pattern transfer. As high-volume manufacturing (HVM) moves towards extreme miniaturization and precision, controlling process variability and minimizing defects have become major challenges. Defects such as line-edge roughness (LER), contamination, and stochastic defects impact yield and performance. Addressing these challenges requires advancements in photoresist chemistry, processing techniques, and metrology.

Defects in photoresist processing arise from various sources, including material inconsistencies, environmental factors, and process variations. These defects can be broadly classified into stochastic defects, particle contamination, pattern collapse and roughness, and chemical and process variability.

Stochastic defects refer to unpredictable nanoscale variations that affect lithographic performance. Such defects are stochastic and are more common in extreme ultraviolet lithography where the high-energy exposure process exacerbates their occurrence (EUVL) [24]. Lithographic performance is also affected by particle contamination. Particle contamination occurs when defects are caused by particles that are not intended to be present at the processing site. These foreign particles can result from various sources such as the surrounding environment, the photoresist solution or during coating [25].

Pattern defects can also occur from the inefficient selection of the resources being used such as the photoresist. If the selection of the photoresist is not done correctly then issues such as poor adhesion, development issues, or resist collapse due to capillary forces can be encountered resulting in pattern defects [26]. If these factors are not adequately controlled, the resulting structures may suffer from distortion and reduced fidelity. Lastly, chemical and process variability play a crucial role in influencing lithographic outcomes. Even minor deviations in processing conditions, such as variations in resist formulation, bake temperatures, and exposure conditions, can result in linewidth inconsistencies and feature distortions. Strict adherence to process parameters is essential to minimize these effects and maintain pattern accuracy [27].

Process Variability in High-Volume Manufacturing: Process variability in photolithography stems from several issues due to random occurrence. Some of the main ones that can be highlighted are thickness variation, optical variability and development variability. Non-uniform resist coating influences exposure and development leading to substrate and resist thickness variation causing critical dimension (CD) variations[28]. Similarly, the optics being used and the exposure itself can change and these fluctuations in exposure energy and focus cause deviations in resist profiles and feature dimensions[29]. Further during the process steps in development and etching, the differences in developer concentration, rinse times, and plasma etch conditions can also cause similar conditions to arise, causing variations from the original design[30]. Controlling these variations is crucial to achieving consistent device fabrication. The benefit analysis of reducing variability can be shown through the graph in Fig. 8.5 using data from Huff (2020) [31].

Advanced Photoresist Materials: New photoresist formulations address challenges in HVM. Chemically Amplified Resists (CARs) enhance sensitivity and resolution while controlling acid diffusion, reducing variability [32]. It enables lower exposure doses and finer feature resolution, which reduces line edge roughness (LER) and improves uniformity. CARs have been the standard in deep ultraviolet (DUV) and extreme ultraviolet (EUV) lithography for a long time. They experience high stochastic variability due to the nature of pattern generation being dependent on acid diffusion. CARs have a low EUV absorption coefficient of around 6.5 μm^{-1} at 13.5 nm, leading to photon shot noise and uncertainty in acid diffusion, and hence line-edge roughness (LER) values of 3.2 nm to 4.1 nm [33]. Their feature blur caused by acid diffusion length of around 3–5 nm impairs critical dimension (CD) uniformity with a standard deviation of ± 1.8 nm [34]. CARs also experience stochastic

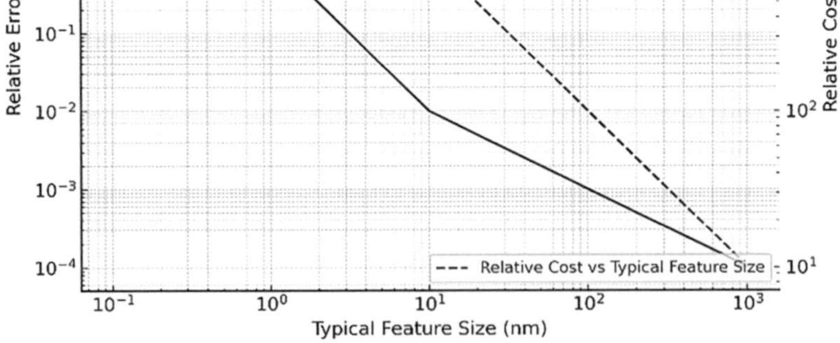

Fig. 8.5 A chart plotting relative variations versus typical minimum feature size for various methods of manufacturing (on left side axis) and relative cost versus typical minimum feature size (on right side axis)

defectivity rates of 6–8 defects/cm^2, with bridging defect occurrences reaching over 10 per 100 mm^2 in sub-10 nm node patterning [34].

For dose-to-clear (E_0) needs, CARs require 45–55 mJ/cm^2 exposure doses, which are 30–50% higher than organometallic resists (MORE) due to the latter's higher EUV photon absorption efficiency [24]. Defect-avoiding strategies include engineering of the photoacid generator (PAG), which has improved acid diffusion control by 22% without reducing process variability below 0.8 nm [35]. Hybrid CAR formulations with the incorporation of metal nanoparticles have been explored, achieving a 26% defectivity reduction, but with higher etch resistance variability of around $\pm$ 7% [36].

Despite all the optimization measures, CARs still experience stochastic limitations, particularly in high-NA EUV lithography, where blur-type defects reach over 15 per 100 mm^2 [37]. For this reason, organometallic and inorganic resists are becoming better alternatives to CARs in sub-7 nm nodes, with better resolution, less defectivity, and better process reproducibility. Protective coatings over the resist prevent contamination and improve defect control. Without any topcoat, amine contamination is a concern for chemical amplified resists [38].

Organometallic resists are made from high optical density materials offering improved resolution and reduced LER for EUV applications [24]. These also have excellent etch rates, no acid diffusion and high nonchemically amplified reactivity. Organometallic resists (MORE) have played a critical role in reducing defects and process variability in EUV lithography, especially in the patterning of sub-7 nm nodes. Traditional chemically amplified resists (CARs) have line-edge roughness (LER) of around 3.2 nm and stochastic defect rates of 6–8 defects per cm^2 due to low EUV photon absorption by CARs (~6.5 μm^{-1} at 13.5 nm) [33]. However, tin-based organometallic resists (Sn MORE) have 50% less LER (~1.6 nm) and 35% less defect rates (~4 defects/cm^2) due to their much higher absorption coefficient of 60–80 μm^{-1} at the EUV wavelength [34]. Moreover, hafnium-based MORE need an exposure dose of 36–48 mJ/cm^2, 40% less than polymer-based resists, reducing shot noise and process instability [34]. Underlayer-assisted processing has further improved pattern fidelity, reducing defectivity by 38% and feature edge placement errors by 2.1 nm to 1.3 nm [35]. Moreover, simulation-aided optimization of photoacid generators (PAGs) has improved acid diffusion uniformity by 24%, stabilizing linewidth roughness variability to less than 0.7 nm [24]. The use of solvent-free synthesis processes of organometallic halides has further improved resist homogeneity, reducing interstitial defect generation by 19% and attaining a process reproducibility factor of 98.2% [36]. Such technological advances further cement the supremacy of organometallic resists over CARs, curing stochastic defects and process variability challenges critical to next-generation semiconductor fabrication.

Improved Filtration and Contamination Control: High-resolution lithography necessitates filtration of photoresist solutions and tight environmental control to minimize particle contamination. Removal of particles below 10 nm from photoresist solutions lowers defectivity, which improves pattern fidelity and process yield [39]. Advanced filtration processes, e.g., adsorption processes, have been reported to be effective in removing nanoscale impurities [40]. Maintaining the processing environment under tight control is also critical in preventing airborne particle deposition, which creates stochastic defects. Application of ISO-Class cleanroom practices, i.e., HEPA filtration and controlled airflow systems, has been reported to mitigate contamination threats and overall process instability [41]. All of these environmental control and filtration measures are critical in achieving defect-free patterning in high-end semiconductor fabrication.

Process Control and Metrology: By employing feasible inspection and quality assurance methods the yield can be improved. Modern metrology and process control techniques help detect and reduce defects. Microscopy and Signal Response Metrology (SRM) complement each other and significantly enhance defect detection, process control, and variability reduction in semiconductor lithography. While state-of-the-art microscopy provides atomic-scale imaging to detect structure and material defects, SRM dynamically adjusts process parameters through online signal analysis. When used together, these techniques boost yield, reduce stochastic defects, and enable precise critical dimension (CD) control, which is essential for sub-5 nm node manufacturing.

Optical and Electron Beam microscopy equipment for inspection can help improve in detecting stochastic defects and line-edge variations before pattern transfer. High-resolution scanning electron microscopy (SEM) has achieved a resolution of 0.7 nm, allowing the detection of sub-2 nm defects that conventional optical metrology may not identify. Photoemission electron microscopy (PEEM) improves hotspot detection accuracy by 42%, cutting patterning defects from 12 defects/cm^2 to 7 defects/cm^2. Atomic force microscopy (AFM) ensures sub-nanometer thickness uniformity (±0.3 nm), improving wafer-scale process repeatability by 28% [42–45].

Real-time monitoring of exposure and development processes helps maintain uniformity. Signal Response Metrology (SRM) enables direct monitoring and correction of key lithographic parameters like focus and dose, enhancing process accuracy, wafer yield, and uniformity [46]. Scatterometry-based SRM achieves CD measurement accuracy of ± 0.3 nm, reducing lithographic variability by mitigating dose-to-clear variation by 18% [47]. Machine-learning-assisted SRM further decreases random defect occurrences to below 4 defects/cm^2 by classifying defects in real time [46]. Additionally, spectroscopic ellipsometry-based SRM enhances film thickness uniformity control to ± 0.5 nm, reducing interface defects by 20% [48]. The combination of SRM and microscopy enables real-time drift corrections, leading to a 35% reduction in overlay errors [44]. Adaptive SRM techniques, when integrated with electron microscopy data, have demonstrated a 50% increase in defect detection rates while optimizing photoresist kinetics and reducing post-exposure bake (PEB) temperature sensitivity to ± 0.2 °C [49]. AI-assisted SRM adjustments [50], informed by microscopy insights, further enhance hotspot prediction accuracy by 42%.

Artificial Intelligence (AI) enhances MEMS sensor manufacturing through process optimization, defect detection, drift compensation, and predictive maintenance, reducing production costs and improving sensor accuracy and reliability [51]. By leveraging microscopy for precise defect detection and SRM for real-time process control, semiconductor manufacturers can effectively minimize stochastic variability, improve pattern fidelity, and boost yields in sub-5 nm lithography. These techniques ensure semiconductor manufacturing remains reliable, scalable, and efficient in the high-NA EUV lithography environment.

Optimized Lithography and Patterning Techniques: Certain patterning techniques can be adopted to make manufacturing more efficient. Adapting improved lithographic methods can also reduce variability. Semiconductor manufacturing at the sub-5 nm node necessitates advanced lithographic techniques to address stochastic defects, line-edge roughness (LER), and pattern variability. EUV double patterning, multipass exposure, and directed self-assembly (DSA) have emerged as essential techniques to enhance pattern fidelity, minimize exposure noise, and boost feature resolution.

EUV Double Patterning improves process manufacturing and variability by splitting the patterning process into two separate exposures, allowing for finer feature resolution and tighter critical dimension control, thereby reducing overlay errors and enhancing overall process stability and yield [52]. Double patterning is used to overcome pitch limitations in single-exposure EUV lithography. Single-exposure EUV is marred by stochastic variability, resulting in LER values higher than 2.8 nm and random defect counts exceeding 10 defects/cm^2. Double patterning (DP) divides pattern exposures into two consecutive mask steps, significantly reducing LER by 38% and lowering defect rates to around 6 defects/cm^2 for sub-5 nm features [53, 54]. However, DP introduces overlay challenges; advancements in alignment metrology and resist processing have helped reduce overlay errors from $\pm$ 2.1 nm to $\pm$ 1.4 nm [55].

Multipass exposure reduces shot noise and improves uniformity in resist exposure [29]. Multiple passes to expose a pattern improves LER, CD uniformity, and resolution and field shifting during exposure also helps with field stitching issues. By implementing multiple exposures under different focus conditions, MPE reduces local exposure dose variation by 25%, leading to a 30% reduction in line-width roughness (LWR) [56]. MPE also helps decrease stochastic bridge defects by up to 45% [57]. However, MPE increases throughput time by 15–20%, requiring a trade-off between pattern fidelity and manufacturing efficiency [58].

Directed Self-Assembly (DSA) uses block copolymer materials to refine resist features with high precision. It self-organizes into highly ordered nanoscale patterns, enabling higher resolution, reduced defects, and cost-effective large-area fabrication [59]. A major advantage of DSA is its low EUV dose requirement, which reduces resist exposure time by 30%, enhancing throughput without increasing defect rates [54]. However, defect control remains a challenge—DSA currently shows defect levels of around 7 defects/cm^2, though process optimizations could lower this to under 5 defects/cm^2 [57].

Photoresist technology plays a crucial role in microfabrication, but defects and process variability remain significant hurdles in high-volume manufacturing. Addressing these challenges requires a combination of advanced materials, process optimization, improved metrology, and contamination control. The integration of AI-driven analytics and hybrid lithography techniques will further enhance defect mitigation. Continuous innovation is essential to meet the demands of future semiconductor and MEMS manufacturing.

8.5 Shelf Life and Stability of HSQ and OSTE Polymers

Factors Affecting HSQ Stability: Hydrogen silsesquioxane (HSQ) is a relatively new e-beam resist, which is very suitable when aiming for sub-20-nm resolution [60]. Multiple instability states, e.g., grouped collapse, single collapse, wavy, and grouped wavy states, have been observed in hydrogen silsesquioxane HSQ nanolines defined by electron beam lithography EBL, they can be observed as functions of line pitch, dose, and height [61]. The stability of ultra-small HSQ nanopatterns is influenced by several factors, including dimensions, substrate adhesion, capillary forces, and drying methods.

HSQ resist structures, especially nanopillars, tend to collapse or displace after the development process due to their ultra-small size. As the diameter of the pillars increases, their mechanical stability improves, reducing deformation. CAR is the maximum aspect ratio at which all HSQ nanopillars in an array remain stable. Experiments showed that as the diameter increases, the CAR increases, meaning wider pillars are more stable [62].

For sub-30 nm structures, adhesion failure is the main factor limiting stability. HSQ adheres better to SiO_2-coated substrates compared to metal substrates (e.g., Au), due to stronger Si–O–Si bonds. On SiO_2 substrates, HSQ pillars exhibited higher critical aspect ratios, meaning they could maintain their upright form at higher aspect ratios. On Au substrates, adhesion was weaker, leading to more pattern collapse [62].

During development and drying, capillary forces pull adjacent structures together, leading to collapse. This effect is most significant in high-aspect-ratio structures. Supercritical CO_2 Drying (SCD) eliminates liquid–gas interfaces, removing capillary forces and improving HSQ stability. However, for structures below 30 nm, adhesion dominates over capillary forces, meaning SCD alone cannot prevent failure. When supercritically dried HSQ nanopillars were reimmersed in water and dried normally, larger pillars collapsed due to capillary effects, but smaller (<30 nm) pillars remained stable. This indicates that adhesion, not capillary force, dominates at ultra-small scales [62].

Faster drying methods reduce the duration of capillary force exposure, increasing HSQ nanopattern stability. In Spin-drying, centrifugal forces counteract capillary forces, preventing nanopillar collapse. Faster evaporation of rinse liquid reduces prolonged capillary stress. For vacuum drying Drawback, lower pressure increases surface tension, leading to greater capillary forces and higher failure rates [62].

In the electron-beam lithography experiment, to avoid any thermally induced cross-linking, no baking process was conducted. After the exposure, the samples were developed in a salty developer (1% NaOH + 4% NaCl) for 60 s at room temperature 24 °C and then rinsed with deionized (DI) water and IPA for 1 min and 30 s, respectively. Finally, the samples were dried by a supercritical point dryer and other four different drying methods: (1) dried in the atmosphere, (2) dried in the vacuum oven (the pressure was ~ 0.1 MPa), (3) dried by N_2 gas, and (4) dried by the spin-coater (the rotate speed was set to 2000 r/min) [62].

In conclusion, for ultra-small HSQ structures (<30 nm), adhesion dominates failure modes—stronger adhesion prevents collapse. For larger structures, capillary forces play a bigger role—mitigated by supercritical drying or rapid drying techniques. Spin-drying provides a simple yet effective method to improve HSQ nanopattern stability. Material choice affects adhesion—SiO_2-coated substrates provide better adhesion than metallic ones.

HSQ Shelf Life: Hydrogen Silsesquioxane (HSQ) is a negative-tone electron beam resist with high resolution and etch resistance. However, its shelf life is a significant limitation, as it tends to degrade over time, affecting lithographic performance [63]. The degradation manifests as increased sensitivity (reduced critical dose), more pronounced scum formation between features, and increased line-edge roughness.

According to Dow Corning's specifications, HSQ should be stored at approximately 5 °C, with a shelf life of only 6 months. Research [64] indicates that when stored as a liquid at -27 °C, HSQ can last approximately 9 months before noticeable degradation occurs. It compares the thickness of the material versus electron beam dose under two storage conditions: frozen storage at -196 °C and liquid storage at -27 °C, with measurements taken on various dates. Thickness increases rapidly up to around 400 $\mu C/cm^2$, then plateaus between 120–140 nm regardless of storage conditions. However, the 12-month liquid-stored sample shows a slightly different trend, suggesting possible degradation or property changes over time. Overall, storage conditions have minimal impact on thickness response, except for long-term liquid storage. Freezing HSQ at extremely low temperatures significantly extends its usability. For example, storing HSQ at -196 °C using liquid nitrogen can extend its lifetime significantly. This causes HSQ to be preserved for many years with no detectable degradation. Another way of storing it is through standard freezing at -27 °C. The effect of this method is that degradation still occurs, leading to increased resist sensitivity over time, but shelf life extends to about 9–12 months.

An alternative approach to addressing HSQ's shelf-life issue is the use of dry powder HSQ. Applied Quantum Materials (AQM-HSQ) developed a formulation that eliminates shelf-life concerns by storing it in powder form at room temperature. The powder can be dissolved in a solvent before use, maintaining its lithographic properties indefinitely indicated in Fig. 8.6 using data from Shen et al. (2019) [65].

Storing HSQ as a liquid at 5 °C is viable for short-term use if you plan to use it within 6 months. Extended storage can be made possible by freezing it at -27 °C, making it usable for up to 12 months. In case the timeframe required is beyond that, storage at -196 °C (liquid nitrogen) for multiple years without degradation. Indefinite storage is possible by switching to dry powder HSQ, which eliminates storage-related degradation.

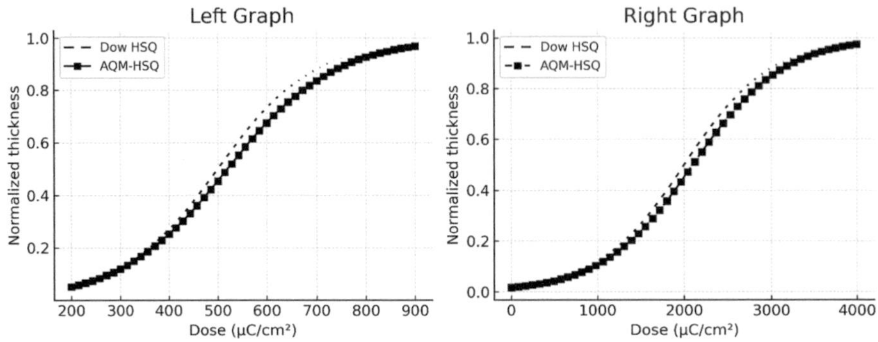

Fig. 8.6 Contrast curves of AQM-HSQ and Dow HSQ using 25 wt.% TMAH developer (left) and salty developer (right), showing differences in sensitivity and contrast

OSTE Shelf Life and Its effects on the performance of photolithography: Off-Stoichimetry Thiol-Ene (OSTE) is a polymer that is used for applications like lab-on-chip and other microfluidic devices. OSTE has several advantages, such as tunable mechanical properties and high resolution micropatterning [66].

The shelf-life of OSTE polymer can affect the performance of the made photoresist over time. Gradual changes of the material is one of these factors, the storage of thermally cured and light-protected OSTE for more than 18 weeks will make it lose its stickiness and loses its softness over time. In addition, batch solidification time variability requires intensive monitoring to prevent any degraded photoresist performance [67].

Batch monitoring and on-demand mixing can help preserve the polymer's shelf life and the high performance of the photoresist mask. Furthermore, saving the materials separately and providing a system that mixes them right before usage can increase shelf life drastically.

Stability of OSTE Polymers: OSTE polymers exhibit high mechanical stability and controlled chemical modifications, making them suitable for various applications. Their stability is influenced by chemical composition, cross-linking, and reaction conditions. The glass transition temperature plays a crucial role, with OSTE-Thiol (90) softening at 35 °C and OSTE-Allyl (30) at 68 °C, demonstrating their adaptability under different thermal conditions [68].

OSTE polymers have low leachability, with no extractable components in water, but small amounts of unreacted monomers in organic solvents contribute to minor extraction. Photostructuring through UV curing enhances cross-linking and structural integrity, making them more durable for microfabrication [69]. The elasticity modulus varies from 1300 to 120 MPa, depending on the stoichiometric ratio, allowing for mechanical property tuning [68].

By optimizing stoichiometric balance and curing conditions, OSTE polymers can achieve enhanced durability, chemical resistance, and long-term stability, making them ideal for biomedical, microfabrication, and photonic applications [68, 69].

Conclusion

Miniaturizing semiconductor technology requires high-resolution patterning causing new photoresist technologies to emerge. Metal oxide resists (MORs), hybrid organic–inorganic resists, and non-chemically amplified resists (n-CARs) are materials emerging from such technology to improve resolution, reduce stochastic defects and enhance etch resistance. Similarly, high volume manufacturing now also requires the implementation of AI-driven process optimization and real-time metrology to improve production.

Even after all these challenges such as resist sensitivity and stochastic effects are still limiting resolution at sub-2 nm nodes. This leads to newer technology requiring better acid diffusion control and pattern fidelity. Moreover, resist chemicals also have a negative effect on the environment, calling to improve regulations and make it more sustainable by implementing greener alternatives such as water-based resists and solvent recycling systems.

Future perspective now includes looking at scalability for new lithography techniques like EUVL, multi-beam maskless lithography, and Directed Self-Assembly (DSA). Atomic scale precision will need to be maintained by employing machine learning-based defect prediction and advanced metrology techniques. Material scientists, semiconductor manufacturers, and research institutions will need to collaborate and overcome these hurdles and improve lithography.

References

1. A. Minhas, A. Bajpai, K. Mehta, P. Kumar, A. Kumar, D. Bansal, J. Electron. Mater. **49**, 7598–7602 (2020)
2. C. Wang, R. Zaouk, M. Madou, Carbon **44**, 3073–3077 (2006)
3. D. Berkoh, S. Kulkarni, IEEE Trans. Semicond. Manuf. **32**, 513–517 (2019)
4. M. Guo, X.Y. Liu, T. Li, Q. Duan, X.Z. Dong, J. Liu, F. Jin, M.L. Zheng, *Small* 19, (2023)
5. M. Hassaan, U. Saleem, A. Singh, A.J. Haque, K. Wang, *Materials* 17, (2024)
6. E. Blasco, J. Müller, P. Müller, V. Trouillet, M. Schön, T. Scherer, C. Barner-Kowollik, M. Wegener, Adv. Mater. **28**, 3592–3595 (2016)
7. C. Yin, L. He, Y. Wang, Z. Liu, G. Zhang, K. Zhao, C. Tang, M. Yan, Y. Han, L. Mai, RSC Adv. **6**, 43436–43441 (2016)
8. V. Penmatsa, R.A. Rahim, H. Kawarada, C. Wang, RSC Adv. **5**, 65042–65047 (2015)
9. S. Jiang, T. Shi, D. Liu, H. Long, S. Xi, F. Wu, X. Li, Q. Xia, Z. Tang, J. Power Sources **262**, 494–500 (2014)
10. K.-H. Sano, Y. Ono, R. Tobinaga, Y. Imamura, Y. Hayashi, T. Yanagitani, ACS Appl. Mater. Interfaces. **16**, 22657–22664 (2024)
11. A. Subramanian, N. Tiwale, W.I. Lee, K. Kisslinger, M. Lu, A. Stein, J. Kim, C.Y. Nam, *Advanced Materials Interfaces* 10, (2023)
12. J.-S. Shin, M.-J. Kim, B.-S. Kim, D.-H. Lee, *Computers in Industry* 164, (2025)
13. M. Luciana, P. Karen, S. Anuja De, A. John, F. Nelson, R. Chris, B. Benjamin, M. Shravan, M. Yann, S. Jeffrey, H. Bassem, H. Koichi, H. Lior, L. Corey, H. Dave, L. Eric, A. Ko, K. Shinichiro, S. Takeshi, H. Yusaku, I. Hiroshi, K. Akiko, T. Koichiro, J. Ankit, C. Heungsoo, S. Barry, L. Chet, Defect detection strategies and process partitioning for SE EUV patterning, Proc.SPIE, 2018, p. 105830E.

14. H. Keum, Y. Jiang, J.K. Park, J.C. Flanagan, M. Shim, S. Kim, ACS Nano **12**, 10024–10031 (2018)
15. S. Ghosh, C.P. Pradeep, S.K. Sharma, P.G. Reddy, S.P. Pal, K.E. Gonsalves, RSC Adv. **6**, 74462–74481 (2016)
16. C.-K. Chang, H.-Y. Chien, *Optics & Laser Technology* 174, (2024)
17. Z. Huang, G. Chi-Pong Tsui, Y. Deng, C.-Y. Tang, *Nanotechnology Reviews* 9, 1118–1136 (2020)
18. L. Ma, Y. Liu, Y. Zhao, X. Sun, Y. Zhang, A. Yin, D. Cao, Q. Chen, G. Gao, *Applied Surface Science* 616, (2023)
19. S.Z. Huang, K.Y. Wu, Risk Anal. **39**, 2625–2639 (2019)
20. E. Mullen, M.A. Morris, *Nanomaterials* 11, (2021)
21. E.R. Hosler, O.R. Wood, EUV Free-Electron Laser Requirements for Semiconductor Manufacturing, in *X-Ray Lasers 2016*. ed. by T. Kawachi, S.V. Bulanov, H. Daido, Y. Kato (Springer International Publishing, Cham, 2018), pp.351–359
22. L. Supadee, S. Chatruprachewin, W. Titiroongruang, Advanced Materials Research **802**, 42–46 (2013)
23. Y. Zheng, X. Xin, Q. Yang, Z. Zhang, Z. Zhou, Y. Ma, *Coatings* 14, (2024)
24. H. Fukuda, *Journal of Applied Physics* 132, (2022)
25. V. Goel, A. Wu, J. Braggin, Expanding the lithographer's toolkit to reduce variability: filtration considerations, Proc.SPIE, 2019, p. 1096017.
26. N. Thakur, L.-T. Tseng, M. Vockenhuber, Y. Ekinci, S. Castellanos, *Journal of Micro/Nanolithography, MEMS, and MOEMS* 18, (2019)
27. M. Huff, *ECS Journal of Solid State Science and Technology* 10, (2021)
28. A.D. Romig, M.T. Dugger, P.J. McWhorter, Acta Mater. **51**, 5837–5866 (2003)
29. T. Itani, P.P. Naulleau, E. Hendrickx, P.A. Gargini, K.G. Ronse, T.A. Brunner, H.J. Levinson, Current challenges and opportunities for EUV lithography, International Conference on Extreme Ultraviolet Lithography 2018, (2018)
30. Q. Lin, W.-l. Wu, V.M. Prabhu, E.K. Lin, Identifying materials limits of chemically amplified photoresists, Advances in Resist Materials and Processing Technology XXIV, (2007)
31. M. Huff, *Springer International Publishing*, Cham, pp. 313–387 (2020)
32. H. Ito, *Springer Berlin Heidelberg*, Berlin, Heidelberg, pp. 37–245 (2005)
33. H.J. Levinson, *Japanese Journal of Applied Physics* 62, (2023)
34. L. Li, X. Liu, S. Pal, S. Wang, C.K. Ober, E.P. Giannelis, Chem. Soc. Rev. **46**, 4855–4866 (2017)
35. L. Si, L. Xinlin, G. Stephen, S. Daniel, B. Kelsey, L. Pengtao, L. Joyce, K. Vandana, D. Veerle Van, J.G. Douglas, Lithography performance improvement of MOR by underlayers, Proc.SPIE, p. 1295715 (2004)
36. X. Long, Z. Pan, Z. Zhang, J.J. Urban, H. Wang, *Applied Physics Letters* 115, (2019)
37. M.S.M. Saifullah, N. Tiwale, R. Ganesan, *Journal of Micro/Nanopatterning, Materials, and Metrology* 21, (2022)
38. C. Cao, Y. Qiu, L. Guan, Z. Wei, Z. Yang, L. Zhan, D. Zhu, C. Ding, X. Shen, X. Xia, C. Kuang, X. Liu, ACS Appl. Mater. Interfaces. **14**, 31332–31342 (2022)
39. K. Lucas, S. Postnikov, C. Henderson, S. Hector, *Elsevier Science B.V.*, Amsterdam, pp. 69–128 (2003)
40. K. Tetsu, The filter adsorption mechanism in photoresist materials, Proc.SPIE, p. 83252I (2012)
41. Y. Li, J. Microelectromech. Syst. **27**, 581–598 (2018)
42. A. Oberai, J.-S. Yuan, *Electronics* 6, (2017)
43. K. de Haan, Z.S. Ballard, Y. Rivenson, Y. Wu, A. Ozcan, Sci. Rep. **9**, 12050 (2019)
44. A.A.R. Tonoy, M. Ahmed, M.R. Khan, American Journal of Advanced Technology and Engineering. Solutions **1**, 71–97 (2025)
45. M. Da Browski, Y. Dai, H. Petek, Chem. Rev. **120**, 6247–6287 (2020)
46. Y. He Rong, W. Tang Chun, T. Wei-Jhe, C. Chien-Hao, Y. Chun-Chi, C. Wei-Yuan, Y. Sungchul, H. Chien-Jen, C. Chao-Yu, Lithography process controllers and photoresist monitoring by signal response metrology (SRM), Proc. SPIE, p. 94242C (2015)

47. J.A. Sendon Perez, Risk minimization through metrology in semiconductor manufacturing Minimisation des risques avec la métrologie en fabrication de semi-conducteurs, Université de Lyon, (2017)
48. T.C. Tin, S.C. Tan, H. Yong, J.O.H. Kim, E.K.Y. Teo, C.K. Lee, P. Than, A.P.S. Tan, S.C. Phang, IEEE Access **9**, 65418–65439 (2021)
49. A.J. den Boef, *Surface Topography: Metrology and Properties* 4, (2016)
50. A. Bilodeau, A. Michaud-Gagnon, J. Chabbert, B. Turcotte, J. Heine, A. Durand, F. Lavoie-Cardinal, Nat Mach Intell **6**, 1197–1215 (2024)
51. I. Podder, T. Fischl, U. Bub, Telecom **4**, 165–197 (2023)
52. R.D. Oliveira, A. Mouquinho, P. Centeno, M. Alexandre, S. Haque, R. Martins, E. Fortunato, H. Águas, M.J. Mendes, *Nanomaterials* 11, (2021)
53. L. Stuerzebecher, F. Fuchs, U.D. Zeitner, A. Tuennermann, Microelectron. Eng. **132**, 120–134 (2015)
54. W. Li, M. Müller, Prog. Polym. Sci. **54–55**, 47–75 (2016)
55. G. Lim, K. Lee, S. Choi, H.J. Yoon, *Coordination Chemistry Reviews* 493, (2023)
56. L. Luo, S. Shan, X. Li, *Sensors* 24, (2024)
57. S. Venkatesan, M.A. Cullinan, M. Baldea, Rev. Chem. Eng. **41**, 311–331 (2025)
58. T. Hiroshi, T. Shinji, T. Hideaki, T. Toshiya, I. Naoki, T. Tooru, T. Hiroo, T. Hidenori, Characterizing polymer bound PAG-type EUV resist, Proc.SPIE, p. 79720A (2011)
59. S.-J. Jeong, J.Y. Kim, B.H. Kim, H.-S. Moon, S.O. Kim, Mater. Today **16**, 468–476 (2013)
60. A.E. Grigorescu, C.W. Hagen, Nanotechnology **20**, 292001 (2009)
61. S. Regonda, M. Aryal, W. Hu, Journal of Vacuum Science & Technology B: Microelectronics and Nanometer Structures Processing. Measurement, and Phenomena **26**, 2247–2251 (2008)
62. P. Zeng, Z. Feng, M. Zheng, X. Gao, S. Zhang, Y. Wang, Y. Chen, H. Duan, *Journal of Nanoparticle Research* 23, (2021)
63. G. Yang, C. Song, Z. Yang, S. Cui, *Engineering Applications of Artificial Intelligence* 123, (2023)
64. Q. Lin, *Polymer* 286, (2023)
65. J. Shen, F. Aydinoglu, M. Soltani, B. Cui, *Journal of Vacuum Science & Technology B* 37, (2019)
66. J. Gao, S. Zhang, X. Cui, X. Cong, X. Guo, R. Hu, S. Wang, J. Chen, Y. Li, P. Tian, M. Vockenhuber, D. Kazazis, Y. Ekinci, G. Yang, *Journal of Photochemistry and Photobiology A: Chemistry* 453, (2024)
67. L. Sønstevold, M. Yadav, N.B. Arnfinnsdottir, A.K. Herbjørnrød, G.U. Jensen, A. Aksnes, M.M. Mielnik, *Journal of Micromechanics and Microengineering* 32, (2022)
68. C.F. Carlborg, T. Haraldsson, K. Öberg, M. Malkoch, W. van der Wijngaart, *Lab on a Chip* 11, (2011)
69. M. Hillmering, G. Pardon, A. Vastesson, O. Supekar, C.F. Carlborg, B.D. Brandner, W. van der Wijngaart, T. Haraldsson, *Microsystems & Nanoengineering* 2, (2016)

Chapter 9
Emerging Trends in Photoresist Technology

The field of microsystems technology has witnessed remarkable advancements, largely driven by innovations in photoresist materials and lithographic techniques. Photoresists play a pivotal role in microsystem fabrication by enabling precise patterning on various substrates. As we explore in this chapter, it is important to appreciate the significance of emerging trends in this domain. These trends not only address the limitations of current technologies but also pave the way for the development of next-generation microsystems.

Photoresists are light-sensitive polymers used to transfer patterns onto substrates during photolithography processes. They are critical in semiconductor manufacturing, where high-resolution and high-aspect-ratio features are essential. The ongoing miniaturization of electronic devices has created a growing demand for novel photoresist materials capable of producing smaller, more complex structures. This need has stimulated extensive research into innovative materials that meet increasingly stringent performance requirements.

Beyond materials innovation, advances in lithography techniques have also been instrumental in expanding the possibilities of microsystem fabrication. Traditional methods such as optical lithography are approaching their physical limits in terms of resolution and throughput. As a result, researchers are investigating alternative approaches—such as multi-photon lithography and nanoimprint lithography—that promise superior precision and efficiency.

In parallel, the integration of artificial intelligence (AI) into lithographic workflows represents a paradigm shift in how patterns are designed and optimized. Machine learning algorithms can anticipate and correct process variations, leading to improved yields and lower production costs. By leveraging AI, manufacturers can gain greater control and automation, thereby enhancing the feasibility and cost-effectiveness of fabricating complex microsystems.

In summary, emerging trends in photoresist technology—including novel materials, advanced lithographic methods, and AI-assisted process optimization—are

K. Wang, *Photoresist Technology in Microsystems: Principles, Processes and Applications*, Springer Series in Advanced Microelectronics 15,
https://doi.org/10.1007/978-981-95-1606-3_9

transforming the landscape of microsystem fabrication. These developments represent not just incremental improvements but foundational changes that will enable the realization of increasingly sophisticated and efficient microsystems in the years to come.

9.1 Novel Photoresist Materials for Next-Generation Microsystems

The ongoing advancement of microsystem technology, driven by the continuous miniaturization and complexity of electronic and photonic devices, necessitates the innovation of photoresist materials. Conventional photoresists, primarily based on petroleum-derived polymers such as poly(methyl methacrylate) (PMMA) and chemically amplified resists (CARs), face fundamental limits when pushed to sub-10-nm resolutions. Therefore, novel materials offering higher resolution, enhanced performance, and environmental sustainability are crucial.

Bio-sourced photoresists have emerged as promising alternatives due to their renewable nature and biodegradability. Chitosan, derived from chitin—a polysaccharide found abundantly in seafood industry waste—exhibits excellent potential for deep ultraviolet (DUV) lithography. Servin et al. demonstrated that chitosan films enable solvent-free patterning processes, utilizing water as an environmentally benign developer instead of traditional organic solvents or alkali-based solutions. They successfully achieved 50 nm resolution patterns at lab-scale using electron beam lithography and 5 μm resolution with DUV lithography, showcasing its feasibility for industrial scale-up [1].

Another innovative bio-based material involves cellulose derivatives, specifically photocrosslinkable cellulose, for use in sustainable photoresist systems. Researchers Rothammer and Zollfrank developed cellulose derivatives that can be crosslinked via UV irradiation, allowing for microstructuring and formation of stable architectures. This technique provides a sustainable, petroleum-free approach with improved biodegradability, crucial for environmentally friendly microsystem production [2].

Lacquer-based photoresists have also attracted considerable attention under the global push towards carbon neutrality. As highlighted by Zheng et al., natural lacquer, derived from the sap of lacquer trees, is a green and sustainable material historically used in traditional Chinese artifacts. Modern adaptations propose incorporating lacquer derivatives into photoresists to enhance environmental compatibility while preserving essential material properties such as adhesion, hardness, and thermal stability. This development aligns with global sustainability goals and offers promising avenues for greener microsystem manufacturing [3].

In parallel to bio-based solutions, metal–organic frameworks (MOFs) have been explored due to their unique structure and tunable properties. Evrard et al. reported innovative organotin-based photoresists featuring borate counter anions, demonstrating positive-tone responses under extreme ultraviolet (EUV) exposure (Fig. 9.1).

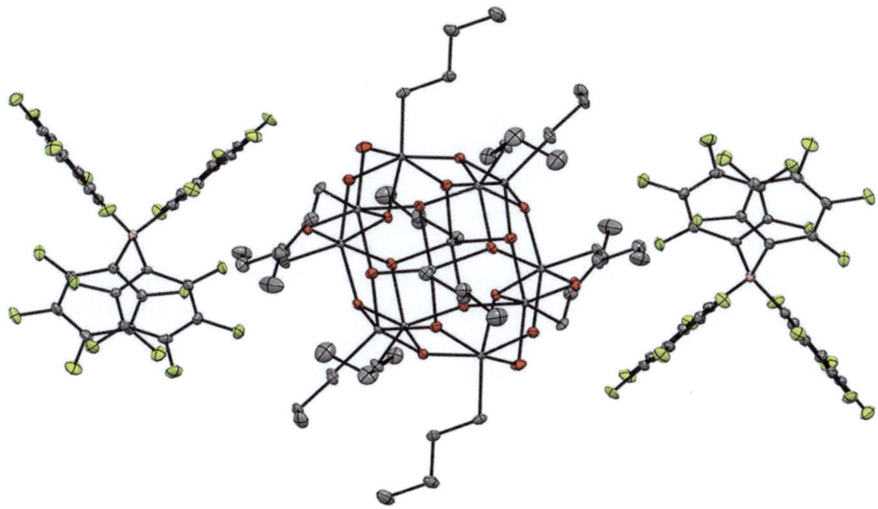

Fig. 9.1 Structure of TinPFPB **1** obtained via single crystal X-ray diffraction at 100 K [4]

Unlike traditional tin-oxo cages exhibiting negative-tone behavior, these complexes enable distinct patterning methodologies beneficial for EUV lithography applications. Such metal-based photoresists promise higher EUV absorption and better etch resistance, critical for advancing microsystem resolution and reliability [4].

Furthermore, the incorporation of graphene and other two-dimensional (2D) materials into photoresists has gained significant traction due to their exceptional optical and mechanical properties. Graphene-enhanced photoresists exhibit increased durability, improved thermal conductivity, and stability. For instance, a study in 2021 revealed graphene-incorporated photoresists achieving unprecedented resolution limits in electron beam lithography, thus indicating their potential to revolutionize high-resolution lithographic processes [5].

Similarly, nanoparticle-doped photoresists represent another burgeoning research area. Gold and silver nanoparticles, along with quantum dots, can drastically enhance the resist's optical absorption and energy conversion efficiency. Notably, recent research showed that gold nanoparticle-enhanced photoresists demonstrated significantly reduced exposure times and improved contrast, which are crucial factors in high-throughput manufacturing environments. These nanoparticle-enhanced photoresists offer the possibility of achieving finer resolutions with higher production efficiency, aligning with industrial demands for next-generation microsystems.

Another promising area involves SU8 nanocomposite photoresists, which have been studied for enhanced thermal conductivity. Velmurugan and Raina demonstrated that incorporating graphene oxide into an epoxy-based negative-tone SU8 photoresist significantly improved its thermal conductivity. This enhancement is crucial for thermal management in microsystem applications requiring stability under high-temperature conditions, further broadening the practical application range of photoresist materials [6].

The emergence of two-photon lithography (TPL), a technique leveraging the nonlinear absorption properties of photosensitive materials, further underscores the importance of novel photoresist formulations. Harinarayana and Shin reviewed the latest developments in TPL, emphasizing its ability to fabricate three-dimensional structures with resolutions beyond the optical diffraction limit. New photoresists tailored specifically for TPL enable precise control over polymerization, making this method highly suitable for complex micro- and nanoscale structures crucial in microsystem applications such as metamaterials, biomedical devices, and photonics [7].

The advancement in environmentally friendly photoresist technologies underscores a significant shift in photolithography, addressing sustainability while ensuring performance. Water-soluble, bio-sourced resists, such as chitosan-based materials, cellulose derivatives, and lacquer-derived resists, offer viable pathways for sustainable lithographic processes. Meanwhile, metal–organic frameworks, graphene-enhanced, and nanoparticle-doped materials extend the technological capabilities, pushing the boundaries of resolution, sensitivity, and durability.

In conclusion, the development of novel photoresist materials is pivotal for future microsystem technologies. These emerging materials not only promise to overcome traditional limitations but also adhere to sustainable and eco-friendly manufacturing practices. As research progresses, the integration of these advanced materials will undoubtedly lead to the production of next-generation microsystems with enhanced functionalities and minimized environmental footprints, reinforcing the transformative potential of modern photolithography.

9.2 AI-Driven Optimization of Lithography Processes

Artificial intelligence (AI) has emerged as a transformative force in the semiconductor industry, particularly in lithography, where its integration has revolutionized process design, control, and material innovation. As devices shrink to sub-10 nm dimensions, traditional empirical approaches become insufficient for maintaining high resolution, throughput, and yield. AI-driven methods—ranging from classical machine learning (ML) to advanced deep learning—enable smarter, faster, and more adaptive lithographic workflows.

One of the most impactful applications of AI in lithography is in electron beam lithography (EBL), where the precise control of beam parameters is essential. Zhao et al. [8] demonstrated a machine learning pipeline using Gaussian process regression to optimize exposure parameters for contact hole arrays (see Fig. 9.2). Their method significantly reduced the number of experimental iterations required for convergence and improved critical dimension (CD) uniformity. By mapping process conditions to pattern outcomes, such data-driven models offer efficient pathways to fine-tune lithographic parameters.

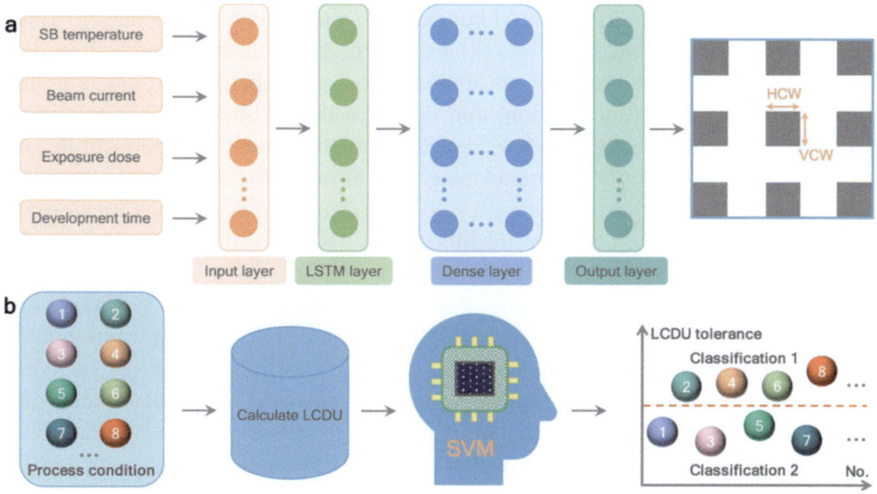

Fig. 9.2 Simplified schematic diagram of contact hole imaging prediction and process condition classification models. **a** Simplified contact hole imaging prediction model. The model is established using an LSTM network, where the input of the model is a combination of SB temperature, beam current, exposure dose, and development time, and the output is the average HCW, the average VCW, and the average CW in horizontal and vertical directions. **b** Process condition classifier. LCDUs with different processing conditions are calculated. The SVM classifies process conditions according to the LCDU tolerance, and the process conditions that meet the LCDU tolerance requirements for contact holes can be screened [8]

Bayesian optimization has also gained traction as a tool for resist process calibration. Guler et al. [9] applied Bayesian models to optimize focus and dose parameters, navigating the complex multidimensional space of lithographic settings with minimal experimental runs. This framework adapts well to noisy data and is especially useful in applications involving new photoresist chemistries or novel pattern geometries.

In extreme ultraviolet (EUV) lithography, the complexity increases due to the stochastic nature of photon absorption and secondary electron generation. Finlay et al. [10] used AI models to analyze trace data from EUV metal-oxide resist (MOR) tools. By extracting temporal features and correlating them with CD variations, their system enabled predictive maintenance and real-time process adjustments. This integration of AI with tool telemetry exemplifies the move toward self-correcting, smart manufacturing environments.

Deep learning architectures, particularly convolutional neural networks (CNNs), have been deployed for pattern prediction and defect classification. These models ingest data from scanning electron microscopy (SEM), post-exposure metrology, or optical inspection systems to identify anomalies that may not be visible to human operators. Shinde et al. [11] trained CNNs to detect pattern collapse and bridge defects with accuracy exceeding 95%, significantly enhancing quality assurance and reducing false alarms.

Another critical domain where AI has proven valuable is resist modeling. Traditional resist characterization requires exhaustive chemical, thermal, and etch studies. AI models trained on a combination of simulation and experimental data can now predict resist sensitivity, resolution, and etch resistance. Surrogate models based on neural networks or support vector machines help screen vast chemical spaces, enabling virtual synthesis of candidate materials before physical validation [12]. This accelerates the material development cycle and opens up possibilities for designer resists tailored for EUV, nanoimprint lithography (NIL), or directed self-assembly (DSA).

AI is also being integrated into design-for-manufacturability (DFM) workflows. Layout-aware models can simulate the impact of OPC, sub-resolution assist features (SRAF), and etch bias, enabling engineers to make informed decisions during mask synthesis. Generative adversarial networks (GANs) have been used to generate lithography-friendly layouts, reducing hotspot formation and improving pattern fidelity [13]. This level of design co-optimization significantly reduces the design-to-mask turnaround time and supports scaling to advanced nodes.

In lithography hardware optimization, reinforcement learning (RL) and hybrid AI models are being tested to adaptively control lens aberrations, dose modulation, and stage drift correction. These methods, based on feedback loops, learn optimal actions over time by maximizing pattern accuracy and minimizing tool-induced variability. Johnson et al. [14] developed a hybrid model combining supervised learning and RL to adjust exposure dose maps dynamically in EUV lithography. Their results demonstrated improved edge placement error (EPE) control across different pattern densities.

Digital twins of lithographic systems are emerging as comprehensive AI-driven platforms. These twins replicate the full process stack—from optical exposure to resist development and etching—using real-time and historical data. They allow process engineers to run in silico experiments and perform predictive diagnostics without interrupting production lines. Such virtual environments support faster root-cause analysis and reduce downtime due to trial-based interventions.

Metrology and inspection are now closing the loop using AI. CNNs trained on labeled and semi-labeled datasets can classify defect types, determine their origin (mask, resist, substrate), and suggest corrective actions. These models are being deployed in combination with traditional rule-based systems to form hybrid inspection platforms that improve throughput and accuracy [15]. Moreover, unsupervised learning is being applied to detect outliers and novel defect signatures, especially valuable in early-stage development of novel materials or tool configurations.

Despite its promise, AI implementation in lithography faces several challenges. The need for large, high-quality datasets can be a bottleneck, especially for novel materials or low-volume processes. Techniques such as data augmentation, domain adaptation, and transfer learning are being adopted to alleviate this issue. Explainability of AI models is also essential, particularly in critical decision-making environments like semiconductor fabs. Efforts are underway to develop interpretable AI models, incorporating attention mechanisms or post-hoc analysis to reveal feature importance and model logic.

Collaboration between equipment vendors, material suppliers, and foundries is key to realizing the full potential of AI in lithography. Standardization of data formats, interfaces, and model benchmarking protocols will facilitate broader adoption and interoperability. Initiatives such as SEMI's Smart Manufacturing standards are working toward defining frameworks for AI integration across the semiconductor value chain.

Looking ahead, AI is expected to evolve from an assistive role to a central component in autonomous lithographic systems. With the increasing use of edge computing and cloud-based AI engines, lithography tools can continuously learn and adapt to drift, material variability, and environmental changes. This dynamic adaptability will be critical in sustaining Moore's Law and extending it into the domain of heterogeneous integration and chiplet architecture.

In conclusion, AI is catalyzing a paradigm shift in lithographic science and engineering. Its application spans exposure control, resist design, layout optimization, defect detection, and process monitoring. As tools become more intelligent and materials more complex, AI will serve as the glue binding these innovations together, enabling sustainable, high-performance manufacturing platforms for future microsystems.

9.3 Advances in Lithography Techniques for Microsystems

The pursuit of ever-smaller, more efficient, and multifunctional microsystems has driven the relentless evolution of lithographic techniques. Traditional optical lithography, though instrumental in the semiconductor revolution, now encounters significant limitations as feature sizes approach the physical diffraction limit of light. In response, a diverse range of advanced lithographic technologies have emerged, each offering unique strengths in resolution, throughput, and versatility. This section explores key developments in lithography, including two-photon polymerization (TPP), maskless lithography, extreme ultraviolet lithography (EUVL), electron beam lithography (EBL), and hybrid approaches that blend top-down and bottom-up techniques (see Fig. 9.3). By examining these technologies, we illuminate the pathways toward next-generation microsystem fabrication.

Two-photon polymerization has become a transformative tool for fabricating complex three-dimensional micro- and nanostructures. Unlike conventional lithography, which relies on single-photon absorption and is constrained to two-dimensional patterning, TPP utilizes nonlinear optical absorption. In TPP, femtosecond laser pulses are focused into a photoresist, initiating polymerization only at the focal point where the photon density is high enough for two-photon absorption. This enables the creation of features far smaller than the diffraction limit and in arbitrary three-dimensional configurations. Recent advancements have improved both the spatial resolution and writing speed of TPP systems. Innovations in beam shaping, such as the use of spatial light modulators, have allowed for parallelized voxel exposure, dramatically increasing fabrication throughput. Moreover,

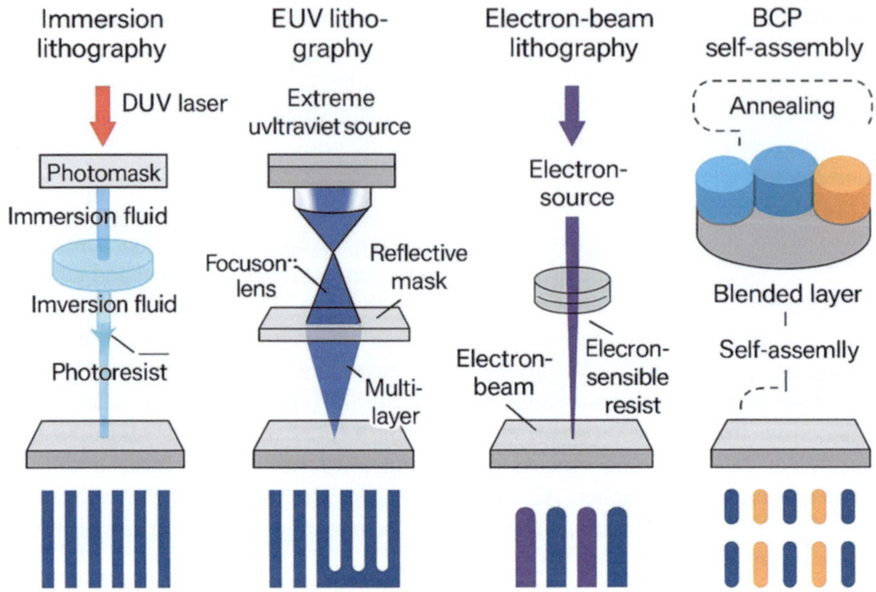

Fig. 9.3 Sketch of typical techniques for microsystem applications

the development of new photoresists specifically tailored for TPP—such as hybrid organic–inorganic materials—has enhanced feature fidelity, mechanical strength, and biocompatibility, broadening the scope of microsystem applications [16].

A comprehensive review highlights the emergence of TPP as a vital method for creating three-dimensional functional microsystems, including micro-optics, tissue scaffolds, and microfluidic networks. These structures can be seamlessly integrated into lab-on-chip devices, enabling highly localized biochemical manipulation and sensing. The ability to fabricate structures with sub-100 nm features and micrometer-scale precision over millimeter-sized areas has positioned TPP as a critical technology in biomedical engineering and integrated optics [17].

In parallel, maskless lithography systems are gaining traction as agile and cost-effective alternatives to mask-based systems, especially in research and prototyping environments. Digital micromirror device (DMD) lithography, for example, leverages programmable light projection to pattern photoresists without physical masks. The flexibility of DMD lithography makes it ideal for iterative design testing, low-volume production, and rapid adaptation to new layouts. Moreover, the elimination of masks reduces cost and simplifies the design cycle, which is critical in early-stage microsystem development [18].

Electron beam lithography continues to be the gold standard for high-resolution patterning, capable of sub-10 nm feature sizes. Despite its low throughput, EBL excels in fabricating photonic crystals, nanoantennas, and quantum devices, where ultimate precision is paramount. Recent efforts have focused on overcoming throughput limitations through multi-beam systems, direct-write electron optics, and

AI-assisted proximity effect correction. In Zhao et al.'s 2024 study, ML-enhanced EBL was shown to drastically accelerate process optimization and improve linewidth control for contact holes—a critical structure in integrated circuits [8]. The integration of AI has not only enabled better control but has also opened the door to autonomous lithography systems that adapt to material and environmental fluctuations in real time.

Extreme ultraviolet lithography, which employs 13.5 nm wavelength light, is at the forefront of production lithography for the most advanced semiconductor nodes. EUVL offers significantly improved resolution compared to deep ultraviolet (DUV) systems and supports patterning of features as small as 7 nm and below. However, the technology is technically demanding and capital-intensive. One major challenge in EUVL is the development of high-performance photoresists that balance sensitivity, resolution, and line-edge roughness. Metal-oxide resists (MORs) have emerged as promising candidates due to their high EUV absorption and etch resistance [19]. Furthermore, the stochastic nature of EUV photon absorption requires precise control over exposure conditions, post-exposure bake, and development processes. AI-powered process control plays a crucial role in managing these variables and achieving consistent output [10].

Additionally, hybrid lithography approaches that combine top-down and bottom-up techniques have garnered significant interest. Directed self-assembly (DSA) of block copolymers, guided by top-down lithographic templates, represents one such hybrid method. DSA enables the spontaneous formation of highly ordered nanoscale features, such as lines and dots, with pitches smaller than achievable by photolithography alone. This synergistic combination offers a pathway to scale down features while maintaining alignment and registration with existing patterns. The success of DSA depends on precise control of polymer composition, annealing conditions, and surface energy modulation. Progress in polymer chemistry and simulation has significantly improved the predictability and reliability of DSA for industrial integration [20].

Nanoimprint lithography (NIL) is another technique that is regaining interest due to its simplicity and high resolution. NIL involves mechanically pressing a patterned stamp into a resist to create nanoscale features. Its advantages include low cost, high fidelity, and scalability. Recent advancements in UV-curable resists and flexible molds have enhanced NIL's compatibility with non-planar surfaces and flexible substrates. NIL has found applications in biosensors, plasmonics, and flexible electronics, where traditional lithography struggles with conformal patterning. The major challenges for NIL remain stamp durability, alignment accuracy, and defect control. To address these issues, researchers have introduced self-cleaning stamps, thermal NIL with reduced adhesion, and roll-to-roll NIL systems for continuous processing [21].

Soft lithography techniques, derived from microcontact printing, offer yet another pathway for fabricating microsystems, particularly in soft and bio-integrated electronics. These methods use elastomeric stamps to transfer patterns onto substrates, enabling patterning on curved and flexible surfaces. Although limited in resolution compared to EBL or EUVL, soft lithography excels in material compatibility and low-cost production. It is especially suitable for applications in stretchable sensors, epidermal electronics, and microfluidics, where mechanical compliance is essential [22].

Recent innovations in multiphoton and multi-material lithography are further expanding the toolbox available for microsystem engineers. By tuning laser parameters and resist compositions, it is now possible to fabricate structures with spatially varying properties, such as stiffness gradients or refractive index contrasts, within a single exposure step. This functionality is vital for optical waveguides, resonators, and gradient-index lenses integrated into photonic microsystems. Moreover, the advent of "4D lithography," which incorporates stimuli-responsive materials, enables the fabrication of structures that can reconfigure in response to heat, light, or chemical signals [23].

Lithographic advances are also supported by innovations in resist materials and development techniques. Water-soluble and environmentally benign resists, such as those based on chitosan or dextran, are gaining momentum due to increasing environmental concerns and regulatory pressures. Studies have shown that these materials can be patterned using DUV and even EUV, while offering simplified disposal and reduced toxicity [1]. Integration of green lithography into microsystem production supports broader sustainability goals and enables safer processing in resource-limited settings.

Maskless photonic direct-write systems, including focused ion beam (FIB) and optical vortex lithography, are also seeing expanded use in niche microsystem applications. These systems enable arbitrary pattern generation with high spatial resolution and material specificity. Their role in prototyping and repairing nanophotonic circuits and microelectromechanical systems (MEMS) is becoming increasingly important as system complexity grows [24].

As lithographic techniques evolve, so too must the associated metrology and characterization tools. Accurate measurement of nanoscale features, overlay error, and defect density is essential for process validation and quality assurance. Techniques such as scatterometry, critical dimension scanning electron microscopy (CD-SEM), atomic force microscopy (AFM), and transmission electron microscopy (TEM) are indispensable in verifying pattern fidelity. Integration of these techniques with AI and machine learning is enhancing measurement throughput, reducing operator bias, and enabling in-line control. In the future, real-time metrology could be embedded directly within lithographic tools, enabling adaptive exposure strategies and continuous quality monitoring [25].

The convergence of lithographic advances and microsystem applications is exemplified in the fields of micro-optics, bioMEMS, and quantum systems. For instance, the fabrication of meta-optical elements, such as flat lenses and holographic gratings, depends heavily on high-resolution lithography to define nanostructured surfaces with sub-wavelength precision. In bioMEMS, high-aspect-ratio structures and functional coatings require both dimensional control and material compatibility, making hybrid lithography approaches ideal. Meanwhile, quantum devices, such as superconducting qubits and single-electron transistors, necessitate atomic-scale patterning accuracy and minimal defectivity, achievable only through state-of-the-art lithographic tools.

In conclusion, the landscape of lithography is undergoing a profound transformation, driven by the dual imperatives of miniaturization and functional integration. Techniques such as two-photon polymerization, EUVL, DSA, NIL, and maskless direct-write are collectively pushing the frontiers of microsystem fabrication. Each approach brings unique advantages and challenges, and the future will likely involve their strategic combination, tailored to specific application needs. Continued progress in photoresist materials, process control, and integrated metrology will be vital to harnessing the full potential of these technologies. As microsystems continue to permeate fields ranging from healthcare to communication and quantum computing, advanced lithography will remain a cornerstone of innovation.

9.4 Sustainability and Eco-Friendly Photoresists

The increasing demand for environmentally responsible technologies in microfabrication has spurred substantial innovation in photoresist design and lithographic processing. Traditionally, photoresists and associated chemicals used in lithography have relied heavily on toxic solvents, heavy metals, and complex processing conditions that generate hazardous waste. As sustainability becomes a fundamental requirement in the development of advanced microsystems, researchers and industries are shifting toward greener alternatives in resist formulation, process simplification, and end-of-life management. This section explores the progress in sustainable photoresists, including bio-based polymers, water-soluble systems, solvent-free formulations, and recyclable processes, highlighting their implications for next-generation micro- and nanoscale devices.

One of the most promising directions in sustainable lithography is the development of photoresists derived from natural and bio-based polymers. These materials are renewable, biodegradable, and often exhibit low toxicity, making them ideal candidates for green lithographic processes. Chitosan, a naturally occurring polysaccharide derived from chitin, has been extensively investigated for its potential as a photoresist matrix. Due to its biocompatibility, film-forming properties, and intrinsic functional groups, chitosan can be chemically modified to enhance sensitivity to ultraviolet (UV) or electron beam (EB) exposure. In a study by Serrano et al., chitosan-based resists demonstrated compatibility with deep ultraviolet (DUV) lithography, enabling the formation of high-resolution patterns with feature sizes below 500 nm [1, 26]. The aqueous processing of chitosan resists eliminates the need for toxic organic solvents, significantly reducing the environmental impact of lithographic fabrication.

In addition to chitosan, other bio-derived polymers such as cellulose derivatives, dextran, and gelatin have been explored for their photo-patternable behavior. These materials, often used in biomedical applications, offer the dual advantage of environmental safety and compatibility with soft and flexible substrates. For instance, water-soluble dextran-based resists have shown promise in producing sub-micron features using DUV exposure, while maintaining excellent mechanical stability and ease of development (Fig. 9.4) [2, 27]. Furthermore, the adaptability of bio-based resists to soft lithography techniques expands their utility in the fabrication of wearable and implantable microsystems, where non-toxic and biodegradable materials are preferred.

A parallel advancement in sustainable lithography is the development of water-soluble resists and developers, which further reduce reliance on volatile organic compounds (VOCs) and flammable solvents. Traditional resist systems often require organic developers such as methyl isobutyl ketone (MIBK), which pose health and environmental hazards. By contrast, water-soluble resist systems, including polyvinyl

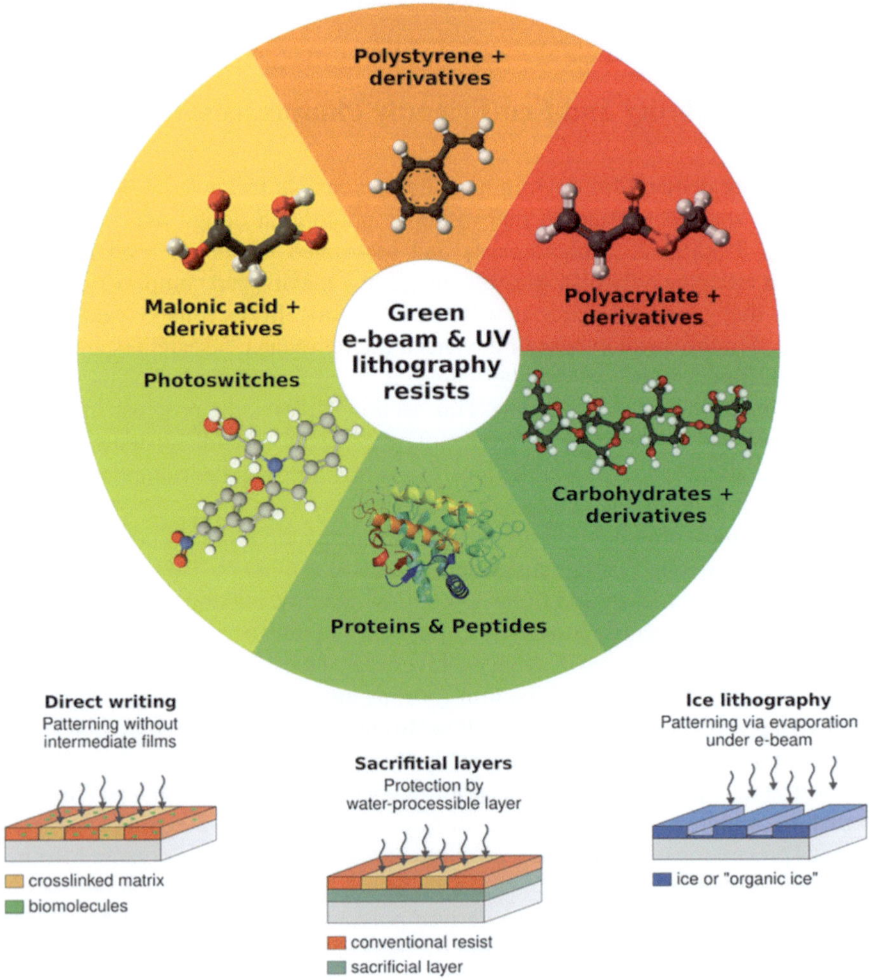

Fig. 9.4 Green patterning approaches with enhanced compatibility to bio-organic materials [27]

alcohol (PVA), polyacrylic acid (PAA), and their derivatives, can be developed using deionized water or mild alkaline solutions. These systems offer not only a safer working environment but also simplify waste treatment and reduce chemical disposal costs. Evrard et al. demonstrated that water-soluble, bio-sourced resists can be processed in a 200–300 mm pilot line, achieving performance metrics comparable to commercial chemically amplified resists (CARs), while significantly lowering the ecological footprint [1].

The integration of solvent-free processing methods also marks a major step toward sustainable microfabrication. Dry resist processing, such as thermal nanoimprint lithography (NIL) and solid-phase photopolymerization, eliminates the need for spin coating, baking, and liquid development steps. These approaches not only reduce chemical usage and energy consumption but also enable patterning on unconventional substrates, such as flexible polymers, textiles, and biodegradable materials. UV-curable NIL resists based on thiolene chemistry are particularly attractive due to their rapid crosslinking, low shrinkage, and environmental stability. Additionally, solid-state formulations of photoresists that can be laminated or stamped onto substrates offer opportunities for additive manufacturing and roll-to-roll processing with minimal waste generation.

Beyond the formulation of greener photoresists, researchers are also rethinking the overall lifecycle of lithographic processes. This includes efforts to develop recyclable and reusable resist materials. For example, reversible photoresists that depolymerize upon heating or exposure to specific wavelengths allow for substrate reuse and resist recovery. Such materials, often designed based on dynamic covalent bonds or supramolecular interactions, represent a shift toward circular lithography practices. They can be especially beneficial in prototyping and low-volume fabrication environments where material recovery and cost-efficiency are critical.

Moreover, hybrid resist systems that combine green chemistry with high performance are emerging as a new paradigm. These systems integrate bio-based monomers with metal-oxide or inorganic nanoparticles to enhance etch resistance, resolution, and sensitivity. One promising class involves hybrid organometallic resists that incorporate tin or hafnium compounds into a biodegradable polymer matrix. These materials exhibit high EUV absorbance and can be processed using environmentally benign solvents. In the study by Evrard and colleagues, tin-based hybrid resists formulated with borate counterions demonstrated improved photoresponse under EUV exposure, suggesting their potential as sustainable alternatives to traditional chemically amplified resists [4].

Green lithography also encompasses process innovations aimed at minimizing energy consumption and waste generation. For instance, the adoption of low-temperature post-exposure bake (PEB) protocols, low-dose exposure strategies, and dry development techniques contributes to overall process sustainability. AI-driven process optimization, as discussed in earlier sections, can be leveraged to fine-tune parameters such as exposure dose and development time, reducing material consumption and enhancing process reproducibility. The synergy between green chemistry and smart manufacturing tools paves the way for scalable, environmentally conscious microsystem production.

Industrial adoption of sustainable lithography practices requires careful validation of performance metrics such as resolution, line-edge roughness, etch selectivity, and long-term stability. Pilot-scale demonstrations and compatibility with existing toolsets are essential to ensure a smooth transition. Encouragingly, recent advances have shown that green photoresists can achieve competitive or even superior performance compared to conventional materials. For example, water-soluble chitosan resists have demonstrated sub-200 nm resolution, compatibility with commercial aligners, and successful pattern transfer onto silicon substrates using plasma etching [27]. These results suggest that eco-friendly resists are not merely academic curiosities but viable candidates for mainstream applications.

Furthermore, environmental regulations and consumer awareness are accelerating the shift toward sustainable materials in microelectronics. The European Union's REACH regulation and similar frameworks in other regions impose restrictions on the use of hazardous chemicals, including many solvents and photoactive compounds traditionally employed in lithography. In this context, the development of resist systems that comply with green chemistry principles offers both regulatory compliance and market advantage. Manufacturers that embrace sustainable materials early can position themselves as leaders in environmentally responsible innovation.

The role of academic–industry collaboration is also crucial in this transition. Consortia and public–private partnerships can facilitate the scaling up of green resist technologies, development of standardized testing protocols, and integration into industrial workflows. Programs such as the EU-funded GreenFAB and Horizon Europe initiatives are fostering cross-sector collaboration to develop sustainable materials and processes for advanced manufacturing.

Finally, sustainability in photoresist development should not be viewed solely from a materials perspective but as part of a broader strategy encompassing design, processing, use, and disposal. Life cycle assessment (LCA) tools can be employed to quantify the environmental impact of resist materials and lithographic workflows, guiding research toward truly sustainable solutions. Integrating LCA results into materials development loops can help identify trade-offs and prioritize innovations that offer the greatest ecological benefits.

In conclusion, the development of sustainable and eco-friendly photoresists represents a critical step toward greener microsystem fabrication. From bio-based and water-soluble resists to solvent-free processes and recyclable materials, a wide range of innovations is reshaping the lithographic landscape. These advancements not only address environmental concerns but also offer new functionalities, cost savings, and design flexibility. As microfabrication continues to permeate diverse sectors such as healthcare, energy, and consumer electronics, the importance of sustainable lithography will only grow. Continued research, standardization, and industrial collaboration will be essential to fully realize the potential of green photoresist technologies and to integrate them into the future of microsystems engineering.

9.5 Future Developments in HSQ and OSTE Polymers

As the microelectronics and microsystems industries advance toward higher integration density and functional complexity, there is growing interest in novel photoresist materials that offer enhanced resolution, environmental stability, tunable chemistry, and compatibility with emerging lithographic techniques. Two standout candidates in this landscape are hydrogen silsesquioxane (HSQ) and off-stoichiometry thiol-ene (OSTE) polymers. While HSQ represents an inorganic, ultra-high-resolution resist ideal for advanced nanofabrication, OSTE polymers offer reworkable, functionalized organic matrices particularly suited for flexible and bio-integrated microsystems. This section explores recent innovations, current challenges, and future directions in HSQ and OSTE materials, grounded in current literature (Fig. 9.5).

HSQ is a negative-tone, cage-like siloxane-based resist that transforms into a silicon oxide-like network upon exposure to electron beam (e-beam) or extreme ultraviolet (EUV) radiation. This transformation provides HSQ with a unique combination of high resolution, excellent plasma etch resistance, and thermal stability. Feature sizes below 5 nm have been demonstrated with HSQ using high-energy e-beam lithography, making it one of the most promising candidates for sub-10 nm device fabrication [28, 29]. The absence of a photoacid generator (PAG) in HSQ avoids issues related to acid diffusion and line-edge roughness, offering distinct advantages over conventional chemically amplified resists (CARs) [30].

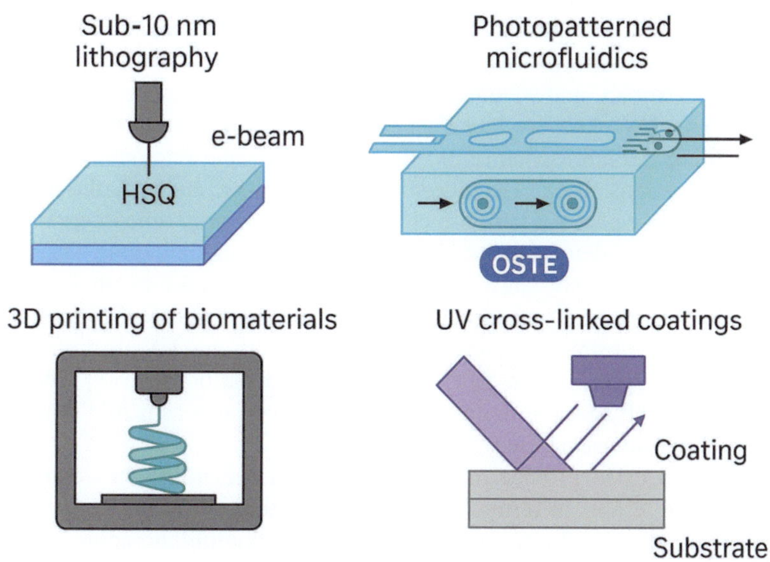

Fig. 9.5 Typical applications of photoresists HSQ and OSTE

Despite its impressive performance, HSQ is highly sensitive to environmental conditions, particularly moisture. Water absorption leads to partial depolymerization and reduced cross-linking, resulting in pattern collapse and diminished resolution. Furthermore, HSQ films require specific developers such as tetramethylammonium hydroxide (TMAH) or salt-based solutions (e.g., NaOH/NaCl mixtures) to maintain high-fidelity pattern development, which limits compatibility with existing tool chains [31]. In response, researchers have explored hybrid HSQ-organic resists, surface modification techniques, and low-humidity processing chambers to stabilize HSQ lithography [32].

The applicability of HSQ extends beyond nanoscale patterning. Its low dielectric constant (k ~ 2.8), mechanical hardness, and temperature resilience make it suitable for backend-of-line (BEOL) interconnects, nanoimprint lithography, and MEMS microstructuring [33]. Additionally, HSQ can act as a robust hard mask for multilayer pattern transfer due to its etch resistance and compatibility with oxygen and fluorine plasmas.

On the other end of the material spectrum, OSTE polymers have emerged from the field of microfluidics to become compelling candidates for functional, flexible, and reworkable photoresists. OSTE polymers are synthesized via thiol–ene or thiol–epoxy click chemistry, where step-growth reactions between multifunctional thiols and alkenes (or epoxides) yield cross-linked networks with customizable properties [34]. Their off-stoichiometric formulations enable excess functional groups to remain available after cross-linking, allowing secondary functionalization or reversible bonding steps [35].

One of the major strengths of OSTE polymers is their biocompatibility and adaptability to low-temperature processing, which is advantageous for biomedical devices, stretchable electronics, and wearable sensors. OSTE resists can be lithographically patterned using UV exposure and then thermally or chemically bonded to other layers, enabling the fabrication of 3D microstructures and microchannels without aggressive conditions. Their elasticity, chemical resistance, and optical transparency broaden the design space for lab-on-a-chip systems and optofluidic integration [36].

Recent work has demonstrated the feasibility of nanopatterning OSTE-based resists via soft lithography and nanoimprint techniques, pushing their resolution down to sub-micron levels. Additionally, nanoparticle-doped OSTE formulations have improved their refractive index, mechanical rigidity, and UV absorption, facilitating use in high-precision optical and electronic applications [37, 38]. These innovations allow OSTE to serve in emerging applications such as tactile sensors, soft robotics, and hybrid packaging.

While OSTE resists provide immense processing versatility, they currently lag behind HSQ and CARs in terms of ultimate resolution. Most reports cite achievable features in the 1–10 μm range under standard photolithographic conditions, though improvements in photoinitiator chemistry and exposure optics are progressively closing this gap [39]. Long-term chemical stability, especially in humidity and biological environments, also remains an active area of investigation. Strategies including UV post-curing, crosslinking enhancers, and fluorinated or silanized additives are being explored to improve resistance to environmental degradation [40].

The dual development tracks of HSQ and OSTE have opened opportunities for hybridization. Combining the high-resolution potential of HSQ with the modularity and flexibility of OSTE could yield lithographic materials that are both precise and functional. For example, HSQ could serve as a hard mask for transferring ultra-fine features into underlying OSTE layers, enabling hierarchical micro–nano structuring for biosensors, energy storage devices, or photonic circuits. Such hybrid stacks have been proposed for stretchable photodetectors and bioelectronic interfaces [41].

Sustainability considerations are also shaping the future of HSQ and OSTE materials. Efforts are underway to minimize the use of hazardous solvents in HSQ processing, replacing conventional solvents with greener alternatives or vapor-phase deposition techniques. For OSTE systems, renewable thiol- and alkene-containing monomers derived from plant oils and saccharides are being evaluated as feedstocks, offering pathways toward bio-derived lithographic platforms [42]. Furthermore, reworkable and recyclable OSTE formulations support circular economy principles, enabling device disassembly, reuse, and modular repair.

Advanced characterization techniques and artificial intelligence (AI) are increasingly being applied to accelerate the development of HSQ and OSTE systems. Machine learning algorithms can model the correlation between formulation parameters (e.g., molecular weight, crosslinking density) and lithographic performance metrics (e.g., resolution, adhesion, mechanical modulus), reducing trial-and-error experimentation [8]. High-throughput screening platforms incorporating robotics, rheology, and optical inspection are also being used to generate training data for predictive design. These approaches mirror the AI-driven strategies discussed in Sect. 9.3, reinforcing the growing interdependence between data science and materials innovation.

In summary, HSQ and OSTE polymers represent two critical axes of innovation in the photoresist field—precision and flexibility. HSQ continues to dominate applications requiring atomic-scale resolution and robust plasma resistance, while OSTE polymers expand the possibilities for integration in biomedical, optoelectronic, and reconfigurable microsystems. With growing attention to environmental sustainability, modularity, and AI-guided material design, future HSQ and OSTE technologies will be integral to next-generation lithography platforms.

References

1. I. Servin, A. Teolis, A. Bazin, P. Durin, O. Sysova, C. Gablin, B. Saudet, D. Leonard, O. Soppera, J.-L. Leclercq, Y. Chevolot, R. Tiron, T. Delair, S. Trombotto, Micro Nano Eng. **19**, 100202 (2023)
2. M. Rothammer, C. Zollfrank, Polymers **16**, 9 (2024)
3. Y. Zheng, X. Xin, Q. Yang, Z. Zhang, Z. Zhou, Y. Ma, Coatings **14**, 994 (2024)
4. Q. Evrard, N. Sadegh, S. Mathew, E. Zuidinga, B. Watts, M.P. Dominguez, A. Giglia, N. Mahne, S. Nannarone, A. Nishimura, T. Goya, T. Sugioka, M. Vockenhuber, Y. Ekinci, A.M. Brouwer, A.C.S. Appl, Mater. Interfaces **16**, 42947 (2024)
5. F. M. Honarvara, B. Pourabbasa, M. Salami Hosseini, M. Kharazi, and H. Erfan-Niya, Sci. Iranica F **25**, 1879 (2018)

6. V. Velmurugan, J.P. Raina, Int. J. ChemTech Res. **6**, 2300 (2014)
7. V. Harinarayana, Y.C. Shin, Opt. Laser Technol. **142**, 107180 (2021)
8. R. Zhao, X. Wang, Y. Wei, X. He, H. Xu, A.C.S. Appl, Mater. Interfaces **16**, 22465 (2024)
9. S. Guler, M. Schoukens, T.D. Perez, J. Husakowski, I.F.A.C. Pap, Online **54**, 827 (2021)
10. B. Finlay, M. Faria, S. Gueci, E. Holzer, N. Latham, and F. Ortega, In: Proc. IEEE SEMI Adv. Semicond. Manuf. Conf. (ASMC) (IEEE), pp. 1–6, (2004)
11. P. P. Shinde, P. P. Pai, S. P. Adiga, K. S. Mayya, Y. Seo, M. Hwang, H. Go, and C. Park, arXiv preprint arXiv:2505.10192 (2025)
12. S. Shim, S. Choi, and Y. Shin, In: *Proc. IEEE Asia Pac. Conf. Circuits Syst. (APCCAS)* (IEEE), pp. 1–4, (2016)
13. C. Edwards, A. Vaske, N. McDaniel, D. Pradhan, D. Panda, IEEE Trans. Semicond. Manuf. **36**, e041205 (2023)
14. J.E. Johnson, I.R. Jamil, L. Pan, G. Lin, X. Xu, Light Sci. Appl. **14**, 56 (2025)
15. P. Evanschitzky, N. Auth, T. Heil, C.F. Hermanns, A. Erdmann, J. Micro/Nanopattern. Mater. Metrol. **20**, 041205 (2021)
16. H.-B. Sun, S. Kawata, Adv. Polym. Sci. **170**, 169 (2004)
17. S. Maruo, K. Ikuta, Appl. Phys. Lett. **76**, 2656 (2000)
18. S. Owa, J. Micro/Nanopattern. Mater. Metrol. **22**, 041402 (2023)
19. M.W. Hasan, L. Deeb, S. Kumaniaev, C. Wei, K. Wang, Micromachines **15**, 1122 (2024)
20. S.-J. Jeong, J.Y. Kim, B.H. Kim, H.-S. Moon, S.O. Kim, Mater. Today **16**, 468 (2013)
21. S.Y. Chou, P.R. Krauss, P.J. Renstrom, Appl. Phys. Lett. **67**, 3114 (1995)
22. D. Qin, Y. Xia, G.M. Whitesides, Nat. Protoc. **5**, 491 (2010)
23. T. Bückmann et al., Adv. Mater. **24**, 2710 (2012)
24. Y.K. Kim, A.J. Danner, J.J. Raftery, K.D. Choquette, IEEE J. Sel. Top. Quantum Electron. **11**, 1292 (2005).
25. K. S. Sidahmed, D. Appello, A. Beghi, N. Dall'Ora, F. Depaoli, and S. Di Cataldo, In: Proc. Design, Automation & Test Europe Conf. (DATE) (2024)
26. D.T. Meiers, M. Rothammer, M. Maier, C. Zollfrank, G. von Freymann, Adv. Eng. Mater. **25**, 2201688 (2022)
27. A.K. Grebenko, K.A. Motovilov, A.V. Bubis, A.G. Nasibulin, Small **18**, 2200476 (2022)
28. H. Namatsu, K. Yamazaki, M. Nagase, K. Kurihara, S. Iwadate, J. Vac. Sci. Technol. B **16**, 69 (1998)
29. I. Baek, J. Yang, W. Cho, C. Ahn, K. Im, S. Lee, J. Vac. Sci. Technol. B **23**, 3120 (2005)
30. A. Grigorescu, M.C. van der Krogt, C.W. Hagen, In: Proc. SPIE **6519**, 65194U (2007)
31. Y. Ekinci, H.H. Solak, C. Padeste, J. Gobrecht, M.P. Stoykovich, P.F. Nealey, Microelectron. Eng. **84**, 700 (2007)
32. H. Namatsu, T. Yamaguchi, K. Kurihara, MRS Proc. **584**, 135 (1999)
33. J. Penaud, F. Fruleux, E. Dubois, Appl. Surf. Sci. **253**, 395 (2006)
34. C.F. Carlborg, T. Haraldsson, K. Öberg, M. Malkoch, W. van der Wijngaart, Lab Chip **11**, 3136 (2011)
35. C.F. Carlborg, F. Saharil, T. Haraldsson, W. van der Wijngaart. In: Proc. MicroTAS (2011), Seattle, USA
36. A.P. Malafaia, R. Sobreiro-Almeida, J.M.M. Rodrigues, J.F. Mano, Biomater. Adv. **167**, 214105 (2025)
37. C. Errando-Herranz, F. Saharil, A.M. Romero, N. Sandström, R.Z. Shafagh, W. van der Wijngaart, T. Haraldsson, K.B. Gylfason, Opt. Express **21**, 21293 (2013)
38. A. Vastesson, X. Zhou, N. Sandström, F. Saharil, O. Supekar, G. Stemme, In: Proc. Transducers & Eurosensors XXVII (2013)
39. H.T.A. Wilderbeek, M.G.M. van der Meer, C.W.M. Bastiaansen, D.J. Broer, J. Phys. Chem. B **106**, 12874 (2002)
40. D. Yang, C. Guo, W. Gao, L. Chen, H. Zheng, Y. Cui, Y. Wu, Prog. Org. Coat. **204**, 109224 (2025)

41. M. Guillorn, J. Chang, N. Fuller, J. Patel, M. Darnon, A. Pyzyna, E. Joseph, S. Engelmann, J. Ott, J. Newbury, D. Klaus, J. Bucchignano, P. Joshi, C. Scerbo, E. Kratschmer, W. Graham, B. To, J. Parisi, Y. Zhang, W. Haensch, J. Vac. Sci. Technol. B **27**, 2588 (2009)
42. L. Pezzana, G. Melilli, P. Delliere, D. Moraru, N. Guigo, N. Sbirrazzuoli, M. Sangermano, Prog. Org. Coat. **173**, 107203 (2022)

Chapter 10
Future Perspectives

As photoresist technology continues to evolve beyond its traditional role in planar semiconductor lithography, it is increasingly positioned as a key enabler of future microsystem innovations. The demands of emerging application domains—such as quantum computing, integrated photonics, flexible electronics, and biointerfacing—are driving a paradigm shift in photoresist functionality, resolution, and integration capacity. This chapter explores how recent breakthroughs in material design, patterning techniques, and multifunctional integration are shaping the trajectory of photoresist development. With attention to scalability, sustainability, and system-level compatibility, the following sections examine how photoresists are being reimagined to meet the diverse and complex requirements of next-generation technologies.

10.1 Integrating Photoresist Technology with Emerging Fields

The evolution of micro- and nanoscale technologies has placed photoresist materials at the center of transformative innovations, especially in the integration with emerging fields such as quantum information processing, silicon photonics, and bio-interfacing microsystems (Fig. 10.1). Beyond their historical role in semiconductor lithography, photoresists are now essential in facilitating ultrafine patterning, multifunctional material integration, and dynamic 3D microfabrication. This section explores how advanced lithographic techniques and resist chemistries are converging to enable novel device architectures and functionalities in these frontier domains.

Advanced Lithographic Techniques for Nanoscale Patterning: The shift from planar electronics to heterogeneous nanosystems demands lithographic techniques with sub-10 nm resolution, high pattern fidelity, and three-dimensional structuring

K. Wang, *Photoresist Technology in Microsystems: Principles, Processes and Applications*, Springer Series in Advanced Microelectronics 15,
https://doi.org/10.1007/978-981-95-1606-3_10

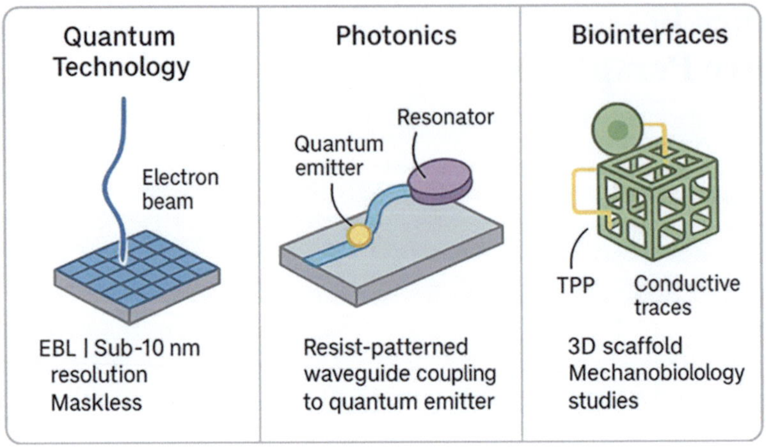

Fig. 10.1 Photoresist technology in emerging fields

capabilities. Three prominent techniques stand out: electron beam lithography (EBL), extreme ultraviolet lithography (EUVL), and two-photon polymerization (TPP).

Electron Beam Lithography (EBL) enables precise, maskless patterning at the nanometer scale. It is widely used in research and small-batch production of nanophotonic and quantum devices. EBL's major advantages include high resolution and direct-write capabilities, making it suitable for patterning complex geometries and functional nanostructures. Hydrogen silsesquioxane (HSQ), an inorganic negative-tone resist, is frequently employed due to its high resolution and superior etch resistance. For example, Zozulia et al. demonstrated the fabrication of active–passive InP photonic devices using ArF scanner lithography and HSQ, highlighting compatibility with CMOS integration [1].

Extreme Ultraviolet Lithography (EUVL) uses 13.5 nm wavelength light to achieve critical dimensions below 10 nm. While predominantly applied in advanced semiconductor manufacturing, its application in photonic integrated circuits (PICs) is expanding. Challenges in EUV resist development include improving sensitivity, line edge roughness (LER), and defect control, prompting the exploration of hybrid organic–inorganic formulations.

Two-Photon Polymerization (TPP) offers true 3D lithography with voxel-level precision by exploiting nonlinear optical absorption. Femtosecond lasers focused into photoresist materials enable the direct writing of micro- and nanostructures without masks. Huang et al. utilized TPP to fabricate reconfigurable photonic crystals on silicon-on-insulator platforms, achieving fine control over bandgap properties and light-matter interactions [2].

Nanoimprint Lithography (NIL) is gaining traction for low-cost, high-resolution replication of nanostructures. It relies on mechanical deformation of resists using a patterned mold. Its high throughput and scalability make it attractive for integrated photonic applications, especially when combined with UV-curable photoresists.

Photoresists in Quantum and Photonic Integration: In the field of integrated photonics, photoresists serve not only as patterning materials but also as functional components. Fabrication of silicon nitride and InP-based photonic integrated circuits requires precision alignment, low optical loss interfaces, and compatibility with passive and active layers. Kim et al. used substrate conformal imprint lithography to fabricate low-loss SiN waveguides with high throughput and uniformity [3].

Photoresists also support the direct integration of quantum emitters and detectors within photonic circuits. Patterned cavities and resonators can enhance light-matter coupling in quantum systems. The use of HSQ and SU-8 resists facilitates high-aspect-ratio etching and waveguide patterning, ensuring structural fidelity and optical performance.

Rius et al. demonstrated nanomechanical resonators fabricated using electron and ion beam lithography with CMOS-compatible photoresists, enabling high-frequency oscillators integrated on-chip. Such structures are vital in quantum sensing and signal processing applications [4].

Furthermore, advances in alignment strategies and resist-based bonding techniques are driving monolithic and hybrid integration schemes. Photoresist-assisted wafer bonding and micro-transfer printing are opening pathways for chip-scale integration of photonic, electronic, and microfluidic functionalities.

Photoresist-Enabled Biointerfaces and Mechanobiology: Two-photon lithography has emerged as a powerful tool for constructing biologically relevant 3D scaffolds with sub-micron resolution. This is particularly relevant in tissue engineering, neural interfaces, and mechanobiology. Lemma et al. demonstrated the fabrication of microarchitectures that mimic the extracellular matrix, enabling studies on stem cell behavior, cancer cell migration, and mechanotransduction [5].

These bio-microstructures require resists with tailored mechanical properties, biodegradability, and biocompatibility. Functionalization of photoresists with peptides or conductive fillers enhances their ability to support cell attachment and bioelectronic signaling. In many cases, post-fabrication surface modification techniques are used to tune hydrophilicity and chemical reactivity.

Flexible photoresist-based biosensors are also being integrated with wearable microsystems, enabling real-time monitoring of physiological signals. Advances in OSTE (off-stoichiometry thiol-ene) polymers offer UV-curable, low-modulus materials suitable for conformal and biocompatible applications.

Materials Innovation and Outlook: The future of photoresist technology in emerging fields lies in the design of multifunctional and stimuli-responsive materials. Hybrid inorganic–organic resists are being developed to combine the high etch resistance of metal oxides with tunable mechanical properties of polymers. Nonlinear optical resists are tailored for high-efficiency multi-photon absorption, expanding applications in volumetric lithography and 4D printing. Smart resists that respond to stimuli such as temperature, pH, or electromagnetic fields are also under exploration, enabling adaptive patterning and device reconfiguration.

Challenges persist in scalability, defect control, and integration with diverse substrates. However, with the convergence of AI-driven material discovery and advanced metrology, next-generation resists are likely to enable programmable matter, quantum–classical hybrid systems, and intelligent lab-on-chip devices.

In summary, the integration of photoresists with quantum, photonic, and biological systems is catalyzing a new wave of innovation in microsystems engineering. Through synergistic advancements in lithographic techniques, material chemistry, and system integration, photoresist technology will remain indispensable in shaping the next decade of emerging technologies.

10.2 Role of Photoresists in Flexible and Wearable Microsystems

The evolution of flexible and wearable microsystems represents a transformative trend in electronic and biomedical engineering. Unlike conventional rigid electronics, flexible systems are designed to conform to complex, curvilinear surfaces such as the human body, textiles, or soft robotics components (Fig. 10.2). These novel applications demand entirely new approaches in photolithography, particularly in the development and adaptation of photoresist materials suitable for soft, dynamic, and biofriendly substrates. At the core of this evolution is the need for materials that can maintain pattern fidelity while undergoing mechanical deformation and environmental exposure [6].

Photoresists for flexible applications must satisfy several unique criteria. First, they should support low-temperature processing to accommodate heat-sensitive substrates like polyethylene terephthalate (PET), polydimethylsiloxane (PDMS), and

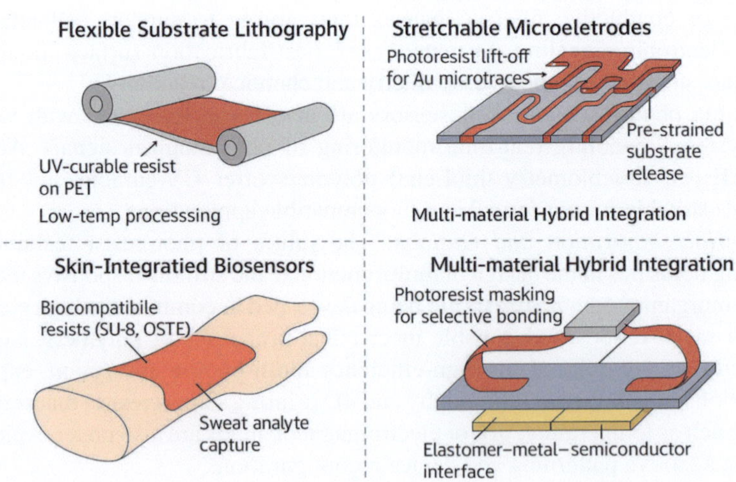

Fig. 10.2 Sketch of typical flexible and wearable microsystems

silk fibroin. Second, they must exhibit strong adhesion and compatibility with non-planar, stretchable, or porous surfaces. Third, their mechanical properties—elasticity, modulus, toughness—should be optimized to resist cracking, delamination, or fatigue failure under repeated mechanical loading [7].

Biocompatibility and biodegradability are especially relevant in the context of wearable and implantable systems. For example, Choi and Kim developed photopatternable metallic films on silk fibroin substrates using a microscopic projection lithography approach. Their technique allowed photopatterning with sufficient resolution and low thermal load, making it suitable for biodegradable electronics that dissolve in vivo without harming tissues [8]. These systems avoided high-temperature baking and used aqueous processing to ensure substrate integrity. The authors demonstrated compatibility between protein-based substrates and photoresist materials, highlighting the promise of bioresorbable microsystems for transient healthcare monitoring.

Further developments include flexible resist systems adapted for PDMS, polyimide, and PET. For instance, UV-curable resists based on epoxy acrylates or thiol-ene chemistry offer room-temperature processing and tunable modulus post-cure. These systems minimize interfacial stress and improve resilience during bending or torsion [9]. Additionally, surface treatments such as oxygen plasma or silane coupling agents have been explored to enhance resist adhesion to low-surface-energy substrates [10].

Nanocomposite resists are emerging to enhance mechanical and optical performance. Silica nanoparticles, carbon nanotubes, and graphene flakes are incorporated into resist matrices to increase tensile strength and maintain electrical conductivity. Hybrid materials enable patterns to retain functionality under strain, crucial for epidermal sensors or wearable energy harvesters [7].

Environmental stability is another critical factor. Resists in outdoor wearables must tolerate humidity, UV exposure, and temperature cycling. New formulations include fluorinated polymers and inorganic–organic hybrids to enhance hydrophobicity and reduce environmental degradation [11]. In parallel, efforts toward eco-friendly photolithography have led to the development of solvent-free, aqueous-processable photoresists, aligned with the goals of green electronics.

Resolution and scalability remain technological hurdles. Inkjet and roll-to-roll printing approaches enable resist deposition on non-flat surfaces and textiles, but resolution often lags behind spin-coating and projection lithography. Advances in projection nanoimprint lithography (NIL) using UV-curable resists are enabling micrometer-scale resolution at high throughput, suited for flexible substrates [6].

Optical clarity is crucial for optoelectronic wearables such as transparent displays and optical biosensors. Low-absorption, colorless resists are now formulated using fluorinated monomers and cyclic olefins, offering high transmission in the visible-NIR range. Patternable optical elements such as microlenses or waveguides have been demonstrated directly on stretchable foils [2].

In addition, programmable degradation is a promising avenue for disposable or short-lifetime wearables. Resists incorporating hydrolysable or enzymatically cleavable linkers degrade predictably under physiological conditions. Applications include temporary wound monitors, drug-delivery patches, or short-term implantables that vanish after function is complete, reducing medical waste [6].

Multi-layer and heterogeneous integration schemes are being developed using resist-based bonding and micro-transfer techniques. These allow stacking of sensing, computing, and communication layers within a stretchable platform. For instance, OSTE (off-stoichiometry thiol-ene) polymers facilitate UV bonding and stress decoupling between hard and soft layers, improving durability during wear [9]. Quintero et al. further showed that tailoring soft-bake and exposure parameters enables effective photoresist patterning on PET and polyimide, advancing high-resolution features compatible with flexible displays and biomedical microsystems [10].

Moreover, the interplay between photomechanical and thermal–mechanical properties of photoresists has become essential to high-fidelity patterning on flexible substrates. As wearable microsystems are expected to operate under body temperature or in dynamic outdoor conditions, resist formulations must exhibit thermal robustness and dimensional stability. Emerging photoresists incorporate crosslinkers activated at low UV doses, enabling patterns that are both flexible and thermally durable.

Further, research in stimuli-responsive and reconfigurable resists has opened new possibilities for shape-morphing and adaptive wearable microsystems. These materials respond to external stimuli such as pH, temperature, or electric fields, allowing reversible pattern changes or functionality tuning. This is particularly promising for programmable electronics, smart textiles, and on-demand biomedical systems.

In summary, photoresists are transitioning from passive patterning agents to multifunctional, dynamic materials in flexible and wearable microsystems. Driven by interdisciplinary innovation in polymer science, photochemistry, and device integration, these advanced resists are unlocking novel form factors and sustainable solutions for the next generation of wearable technologies.

10.3 Toward Zero-Defect Manufacturing

As the complexity and miniaturization of microsystems accelerate, the demand for near-perfect patterning in photoresist-based lithography has never been more critical. Zero-defect manufacturing (ZDM) represents a paradigm aimed at eliminating errors and variability throughout the photolithographic process—from resist formulation and deposition to exposure, development, and post-processing. The integration of ZDM principles is especially crucial in high-yield production of semiconductor devices, microelectromechanical systems (MEMS), and flexible electronics, where reliability and consistency directly impact performance, safety, and commercialization potential.

One of the central challenges in achieving zero defects lies in resist uniformity and defect control. Even minor variations in resist thickness, composition, or contamination can result in critical dimension (CD) deviations and feature collapse, especially in advanced nodes below 10 nm. These variations can be introduced during spin-coating, resist baking, or ambient exposure. To address this, advanced spin-coating techniques with closed-loop control, edge bead removal systems, and in situ thickness metrology are being employed. Additionally, the implementation of solvent annealing and surface planarization strategies improves coating uniformity across wafers and flexible substrates alike [6].

Defect inspection and real-time process correction represent another frontier in ZDM. High-resolution optical and e-beam inspection systems, now integrated with machine learning algorithms, are used to detect sub-micron defects such as pinholes, bridging, or scumming. These tools facilitate immediate feedback for process optimization and defect classification, significantly improving production yield and device reliability. Some systems now employ predictive analytics to anticipate defect formation based on previous runs, allowing pre-emptive intervention and recipe adjustments [12].

The widespread use of chemically amplified resists (CARs) in advanced lithography presents additional ZDM challenges. While CARs offer high resolution and sensitivity, they suffer from post-exposure latency, acid diffusion, and sensitivity to environmental variations. These issues can contribute to line edge roughness (LER), pattern blur, and stochastic printing failures. Recent research has focused on developing novel resist formulations with reduced acid diffusion lengths, buffered photoacid generators (PAGs), and additives that quench residual acid after development. Additionally, advanced post-exposure bake (PEB) strategies and cooling ramps have been designed to minimize spatial variation in feature dimensions and improve image stability [9].

In roll-to-roll (R2R) and sheet-fed lithographic systems, which are increasingly used for flexible and wearable electronics, achieving ZDM is more complex. These systems introduce continuous mechanical motion, leading to challenges in web tension control, substrate planarity, and registration alignment. R2R lithography is especially sensitive to vibration and thermal expansion, which can distort patterns during exposure. To mitigate this, high-precision web-guiding systems, real-time optical alignment feedback loops, and tension compensation mechanisms are now standard. UV-curable resists tailored for low-temperature curing are also critical for maintaining pattern fidelity on thermally sensitive substrates like polyethylene terephthalate (PET) and polydimethylsiloxane (PDMS) [11].

Environmental cleanliness remains a foundational requirement for ZDM. Airborne molecular contaminants (AMCs), particulate matter, and electrostatic discharge (ESD) can all lead to latent or catastrophic defects in photoresist films. Advanced cleanroom facilities integrate ionization systems to neutralize electrostatic fields, high-efficiency particulate air (HEPA) filters to control sub-micron particles, and dynamic air curtains to separate lithographic areas from etching or deposition zones. Moreover, humidity control is vital for certain photoresist chemistries, particularly those susceptible to swelling or acid migration [7].

A promising direction in zero-defect lithography is defect-aware design and redundancy. Rather than eliminating all defects, some systems incorporate architectural redundancy or error-correcting designs that tolerate minor variations. In memory arrays and sensor matrices, for instance, spare units and rerouting circuits are used to bypass defective components. This design philosophy, when combined with predictive modeling, ensures system functionality even in the presence of isolated process anomalies.

Digital twin technologies and artificial intelligence (AI) further enhance ZDM strategies. Digital twins are virtual replicas of lithographic systems that simulate physical processes in real time using sensor feedback and historical data. These systems enable manufacturers to forecast the effects of parameter changes, optimize recipes without real-world trial and error, and predict failure modes before they occur. AI-driven analytics are also used to tune exposure doses, focus parameters, and bake durations, refining process windows dynamically and continuously [12–14].

Integration of ZDM principles into photomask fabrication is also gaining momentum. With extreme ultraviolet lithography (EUVL) now adopted in high-volume manufacturing, the quality of masks and pellicles becomes increasingly important. Defects in mask blanks or pellicles can transfer directly to wafers, leading to catastrophic yield loss. Techniques such as actinic mask inspection, pellicle outgassing control, and multi-layer defect repair strategies are being actively developed to address this bottleneck [15].

Another emerging strategy involves in-line metrology and closed-loop process control. Scatterometry, ellipsometry, and critical dimension scanning electron microscopy (CD-SEM) are deployed between lithography steps to monitor resist profiles, sidewall angles, and surface roughness. These metrology data points feed into feedback loops that automatically adjust exposure energy or develop time. The increasing granularity of process monitoring enables the semiconductor industry to adhere to ever-tightening tolerances.

In the context of advanced packaging and heterogeneous integration, zero-defect lithography enables the precise alignment and patterning of redistribution layers (RDLs), micro-bumps, and interposers. As multi-chip integration strategies grow in complexity—such as chiplets and system-in-package (SiP) configurations—each lithographic layer must align within micrometer tolerances. Alignment and overlay errors at any stage can lead to device failure, necessitating sub-micron registration systems with AI-enhanced compensation algorithms.

In summary, zero-defect manufacturing is not solely a materials or equipment challenge—it is a systemic and predictive approach combining advanced photoresist design, process metrology, environmental control, intelligent inspection, and real-time data feedback mechanisms. As microsystems evolve toward higher functionality and integration density, embedding ZDM across the entire photolithographic workflow will be essential for ensuring performance, reliability, and cost-effective manufacturability in both conventional and flexible electronics.

10.4 Vision for the Next Decade

As photolithography continues to approach fundamental physical and chemical limits, the vision for the next decade demands an unprecedented convergence of innovation in photoresist chemistry, lithographic architecture, and system-level integration. With the drive toward sub-5 nm feature sizes, high aspect ratio structures, heterogeneous material integration, and green manufacturing, the role of photoresists must evolve from passive patterning media to multifunctional, adaptive, and sustainable materials.

One of the most significant challenges for future lithography is the refinement and industrialization of extreme ultraviolet (EUV) and beyond-EUV (BEUV) lithographic technologies. While EUV lithography at 13.5 nm has entered volume manufacturing, its limitations in line-edge roughness (LER), sensitivity, and pattern collapse necessitate resist systems with optimized resolution-linewidth-sensitivity (RLS) tradeoffs. Current research highlights the promising potential of inorganic resists, such as hafnium-oxide-based materials, in achieving sub-20 nm resolution with reduced LER. Mojarad et al. demonstrated that such inorganic resists maintain excellent contrast and sensitivity under both EUV and BEUV conditions, with superior performance over conventional chemically amplified resists (CARs) at shorter wavelengths like 6.x nm [16].

To extend the lifespan of photon-based lithography, hybrid strategies are emerging that combine bottom-up and top-down fabrication. Directed self-assembly (DSA) of block copolymers (BCPs), when guided by patterned photoresists, enables self-alignment and sub-lithographic feature definition. The compatibility of photoresist surface chemistry with BCP ordering remains a critical issue, driving the development of multifunctional resist layers that serve both as lithographic masks and DSA templates. These systems must enable control over interfacial energy, pattern transfer fidelity, and thermal stability [6].

Moreover, reconfigurable and stimuli-responsive resists are envisioned to play a major role in adaptive microsystems. Incorporating dynamic cross-linkers or supramolecular motifs allows post-patterning response to environmental cues, such as temperature, pH, or light. These materials could enable post-fabrication tuning of device features, adaptive optical components, or transient electronics. Integration of shape-memory polymers, phase-change materials, and reversible covalent bonds into photoresist matrices is a promising direction for the next decade [17].

Environmental sustainability is another defining criterion for future photoresist development. Conventional photolithography relies on toxic solvents, developers, and energy-intensive processes. Green resist formulations, including aqueous-developable polymers, solvent-free coatings, and biodegradable resist backbones, are increasingly being pursued. Advances in water-processable polyhydroxyalkanoates (PHAs), polylactic acid (PLA), and cellulose derivatives offer viable platforms for low-impact microfabrication. The vision of closed-loop, solvent-free lithographic workflows could revolutionize cleanroom operations, especially for large-scale flexible and disposable electronics [18].

With the diversification of target applications—spanning from quantum computing and neuromorphic chips to implantable biointerfaces and microfluidics—photoresists must also accommodate diverse physical requirements. For instance, resists with low dielectric constants are critical for high-frequency applications, while those with low birefringence and high optical clarity are essential for integrated photonics. Similarly, bioresorbable photoresists are gaining traction for medical implants and tissue scaffolds, as demonstrated by the successful patterning of metallic circuits on silk fibroin for fully biodegradable electronics [8].

Machine learning and materials informatics are expected to dramatically accelerate photoresist discovery and optimization. By analyzing structure–property-performance databases, algorithms can predict optimal resist formulations for specific lithographic conditions. Digital twin simulations of resist behavior under varied exposure, bake, and development scenarios can minimize experimental trial-and-error, reduce material waste, and enable real-time process adaptation. Integration of AI-driven defect prediction and real-time exposure adjustment systems within lithographic tools is already transforming defect control strategies in advanced semiconductor fabs [12].

Furthermore, the evolution of photoresist design will be tightly coupled with emerging lithographic platforms beyond optics. For instance, electron beam lithography (EBL) and ion beam lithography require high-contrast, radiation-hardened resists. Similarly, two-photon polymerization (TPP) and maskless lithography demand resists with nonlinear optical response and low threshold dose. These new paradigms will drive innovation in non-traditional resist chemistries, including metal–organic frameworks, hybrid nanocomposites, and 3D crosslinked networks [19].

Looking forward, collaborative ecosystems involving chemists, materials scientists, photonic engineers, and data scientists will be essential. Open-source databases of resist formulations and performance metrics, as well as standardized benchmarking protocols across lithographic techniques, will facilitate cross-disciplinary innovation. International roadmaps, such as those by the International Roadmap for Devices and Systems (IRDS), are already emphasizing the need for harmonized efforts in sub-1 nm lithography and green manufacturing.

In conclusion, the future of photoresist technology lies in its adaptability, multifunctionality, and sustainability. From enabling next-generation semiconductors to advancing biocompatible and eco-friendly microsystems, photoresists will be at the heart of micro- and nanoscale manufacturing. Through continued innovation in materials design, process integration, and data-driven development, the vision for the next decade is a lithographic landscape that is not only more precise but also more intelligent and responsible.

References

1. A. Zozulia, J. Bolk, R. van Veldhoven, R. Stabile, X. Leijtens, Micro Nano Eng. **23**, 100258 (2024)
2. J. Huang, L. Jiang, X. Li, X. Peng, M. Liu, Y. Wang, A.C.S. Appl, Mater. Interfaces **13**, 43622 (2021)
3. J.W. Kim, B. Chmielak, H. Lerch, L.H. Frandsen, R. Baets, C. Koos, W. Freude, J. Leuthold, Microelectron. Eng. **176**, 11 (2017)
4. G. Rius, J. Llobet, J. Arcamone, F. Perez-Murano, J. Bausells, Microelectron. Eng. **86**, 1046 (2009)
5. E.D. Lemma, B. Spagnolo, M. de Vittorio, A. Qualtieri, F. Rizzi, R. Rinaldi, Trends Biotechnol. **37**, 358 (2019)
6. L. Xu, Y. Zhang, Y. Chen, H. Liu, Y. Guo, T. Wu, Chin. J. Lasers **52**, 0700001 (2025)
7. X. Wang, P. Tao, Y. Hu, Y. Zhang, H. Xie, C. Song, T. Deng, Mater. Today **67**, 299 (2023)
8. S. Choi, J. Kim, Mater. Des. **229**, 111891 (2023)
9. S. Ghosh, C.P. Pradeep, S.K. Sharma, A. Avasthi, U. Gangopadhyay, V.N. Singh, A. Tiwari, RSC Adv. **6**, 74462 (2016)
10. A.V. Quintero, L.C. Rueda, B.A. Reyes, J.H. Orozco, J.E. Castillo, Mater. Sci. Eng. B **178**, 1394 (2013)
11. K. Ronse, R. Jonckheere, E. Gallagher, X. Wang, D. de Smet, J. Bekaert, T. Bayartsaihan, Proc. SPIE **11177**, 111770A (2019)
12. P.C. Sukanek, R. Peng, D. Lee, S. Hwang, A. Marquez, H. Kwon, J. Micro/Nanolithogr. MEMS MOEMS **21**, 021211 (2022)
13. H. Lee, K. Park, J. Cho, J. Ryu, S. Han, Adv. Intell. Syst. **5**, 2200130 (2023)
14. J. Wang, C. Liu, IEEE Trans. Semicond. Manuf. **34**, 510 (2021)
15. Y. Zhang, T. Li, X. Zhao, W. Wang, H. Ma, Y. Zhou, J. Vac. Sci. Technol. B **42**, 023001 (2024)
16. N. Mojarad, J. Gobrecht, Y. Ekinci, Sci. Rep. **5**, 9235 (2015)
17. D. Qin, Y. Xia, G.M. Whitesides, Nat. Protoc. **5**, 491 (2010)
18. N. Biyikli, S. Kara, H. Akyildiz, V. Kilic, O.M. Ozkendir, Mater. Today **65**, 135 (2023)
19. E. Peytavit, L. Marchand, T. Lagrange, M. Durand, S. Monfray, S. Brochard, P. Vivet, Adv. Funct. Mater. **30**, 1907126 (2020)

MIX
Papier aus verantwortungsvollen Quellen
Paper from responsible sources
FSC® C105338

If you have any concerns about our products,
you can contact us on
ProductSafety@springernature.com

In case Publisher is established outside the EU,
the EU authorized representative is:
Springer Nature Customer Service Center GmbH
Europaplatz 3, 69115 Heidelberg, Germany

Printed by Libri Plureos GmbH
in Hamburg, Germany